Plane Trigonometry

Plane Trigonometry

Dedicated to my team

astrarka

Plane Trigonometry

ISBN-13: 978-1466327368

ISBN-10: 1466327367
First Edition

Foreword

We wanted to present the topics in Plane Trigonometry as a set of Lego blocks – conceptual building blocks, each sitting on top of the other. We decided to create layers of solved examples and problems in between meaningful subsets of concepts. This stratification helped us to string together a book, which we hope will help build strong conceptual foundation for the students of High School Mathematics.

The word trigonometry comes from tri "three" + gonia "angle" + metron "a measure", which is a branch of mathematics that deals with relations between sides and angles of triangles. Therefore trigonometry literally translates to "triangle measurement". The primary application of trigonemetry was in heights and distances - and astronomy. Presumably during the second half of the second century B.C., the first trigonometric table was compiled by the astronomer Hipparchus of Nicaea, who thus earned the right to be known as "the father of trigonometry". Systematic study of trigonometric functions reached India as part of Hellenistic astronomy. In Indian astronomy, the study of trigonometric functions flowered in the Gupta period, especially due to Aryabhata. During the Middle Ages, the study of trigonometry was continued in Islamic mathematics, whence it was adopted as a separate subject in the Latin West beginning in the Renaissance with Regiomontanus. The development of modern trigonometry can be traced to the western Age of Enlightenment, beginning with 17th century mathematics and reaching its modern form with Leonhard Euler.

We sincerely hope that the student is able to get a good grasp of the subject and the techniques after working with the content of this book. If the experience of going through this work is joyful for the student and works as a tool for building his / her understanding, we would be satisfied that we have met the primary objective of this effort.

Chandramouli Mahadevan.

Astrarka Educational Solutions Private Limited.

Bangalore, India.

Preface

This work is organized a bit differently from the others. The subject of Plane Trigonometry has been decomposed into a set of concepts. These concepts commence with the notion of an angle and a triangle; and it is on these two concepts the entire domain of Plane Trigonometry stands.

Each concept has been split into a set of axioms and postulates, theorems which require proof or predicates or statements that follow from the knowledge uncovered so far.

After a set of concepts are covered, a set of solved problems are presented which make use of the concepts covered so far. This demonstrates the basic problem solving strategy the student might want to internalize or understand which dealing with problems and challenges in the subject.

The set of solved problems are followed by a set of problems, which the student must try to solve by himself. In doing so, the depth of understanding in the subject improves. Mathematics is not a spectator sport. It requires patience, perseverance and practice. The level of expertise in the subject in some sense is directly proprotional to the number of problems solved by the student. The term "solved" is used to imply accuracy of thought, stringing together intermediate steps and accuracy of the final result. In a way, this term refers to the quality of the means and the quality of the end goal for each problem.

There may be situations where the student is stuck and requires a gentle push to make progress. When the student faces such a deadlock, the helping hand comes in the form of the second part of the book, where all the problems are solved completely.

This work is a comprehensive self study guide for the students who desire to improve their understanding, appearing for Math related competitive examinations and tests.

I believe that Astrarka has been blessed to have had the opportunity to work with some of the best and brightest Any work of this magnitude is always a product of teamwork. R Balasubramanian, Shilpa Jaikumar and Venkatratnam Pandit have contributed a great deal to this effort. A big thanks goes to the family members of our team. They have been a great source of inspiration during this entire effort. They have made a personal sacrifice to ensure that Astrarka succeeds. Without the unflinching commitment and single minded dedication of my team and the members of their family, this book would have been an exercise in futility.

Chandramouli Mahadevan

Table of Contents

1   Introduction 1

2   Good Habits..3

3   Measurement of Angles.......................................5

4   Basic Trigonometric Ratios 13

5   Trigonometric Ratios across the 4 quadrants.....31

6   Trigonometric functions of angles of any size and sign.....39

7   General Expressions for all Trigonometric Ratios 51

8   Trigonometric Ratio: Sum and difference of two angles 57

9   Trigonometric Ratio: Multiple and Sub-multiple angles.....65

10   Solutions of triangles 73

11   Heights and Distances...83

12   Inverse Circular Functions.................................85

13   Problems 89

14   Closing Thoughts 143

1 Introduction

To say that Trigonometry is useful, therefore, we must learn it, is an understatement. This book serves as a conceptual introduction to the subject. It also focuses on problem solving strategies. Any book of Mathematics is incomplete without a bunch of problems to solve. This book is no different. We have organized the material into a sequence of concepts and exercises that make use of those concepts. Some familiarity with algebra, biometry is a prerequisite. Most of the material uses the Pythagorean Theorem. Similarity and congruence of triangles and related theorems and concepts would be extremely handy.

This book must not be read like a work of fiction. Instead, the student is advised to spend quality time in ensuring conceptual understanding. Solving problems in order to verify our conceptual understanding is extremely important. Most of us believe arriving at the final answer is the ultimate goal. We have come across several books on the subject, where the authors have skipped several steps and simply used the phrase "it follows from the fundamental principles ..." and made a conclusion. We disagree with this approach. The purpose of the problem solving is build the path to the solution using first principles or well-known formulas - and build an airtight reasoning on how the problem solving process moves towards the final answer. This serves as a demonstration of our understanding of the subject - basics, formulas and methods of manipulation.

2   Good Habits

There are five fundamental principles, or say good habits that we would like to emphasize before we commence our discussion on Mathematics.

1. Neatness is conducive to accuracy. Refrain from the temptation to write down something quickly and then scratch the same to make the necessary corrections.

2. One of the weaknesses we find in students while solving word problems is the usage of = sign. This sign has a specific meaning in the world of mathematics. It cannot be used as a way to begin every new line or step in the problem solving process. Use appropriate mathematical signs and symbols. Never use them to mean something vague. = Sign is never a good space filler.

3. Spend a second or two to explain how you arrived at a certain step. Several books and references use a statement, such as ``it follows from the above statement''. We have oftentimes wondered how the expression or equation below follows from the one above. A good explanation is an excellent demonstration of your understanding of the underlying principles.

4. When you are faced with several conclusions during a problem solving process, it is a good idea to number the statements or equations. In subsequent steps, you can refer to these conclusions by using the label or the assigned equation number.

5. The easiest of problems attracts the silliest of mistakes. If the problem is easy, motivate yourself to get it right. Do not let overconfidence or carelessness take control of the situation.

3 Measurement of Angles

1. When two line intersect, they form an angle.
2. The two lines are called arms of the angle.
3. The point of intersection is called the vertex.

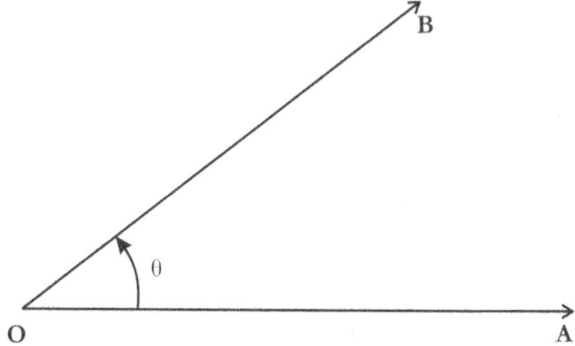

θ : Angle

O : Vertex

$\overline{OA}, \overline{OB}$: Arms of the Angle

4. We come across various units associated with angles - degrees, radians etc. These are units of measurement of angles.
5. We have several systems of measurement of length, namely - centimeters and meters, feet and miles etc. Similarly, we have three major systems for measurement of angles.
6. They are:
    a. Sexagesimal Measure
    b. Centesimal Measure
    c. Circular Measure
7. In the sections that follow, we will get a better understanding of these three systems of measurement. We will also explore ways of converting angular measurements from one system to another.

3.1 Sexagesimal Measure

1. A right angle is divided into 90 equal parts.
2. Each of these parts consitutes a degree. Therefore, 90 degrees make a right angle.
3. Each degree is further divided into 60 equal parts, each part denotes a minute.
4. Each minute is further divided into 60 equal parts, each part denotes a second.

Unit	equals	is denoted by
1 right angle	90 degrees	90^0
1 degree	60 minutes	$60^{'}$
1 minute	60 seconds	$60^{''}$

3.2 Centesimal Measure

1. A right angle is divided into 100 equal parts.
2. Each of these parts consitutes a grade. Therefore, 100 degrees make a right angle.
3. Each grade is further divided into 100 equal parts, each part denotes a minute.
4. Each minute is further divided into 100 equal parts, each part denotes a second.

Unit	equals	is denoted by
1 right angle	100 grades	90^g
1 grade	100 minutes	$100^{'}$
1 minute	100 seconds	$100^{''}$

3.3 Conversion from Sexagesimal to Centesimal measures

1. 1 right angle $= 1^0 = 100^g$

2. $1^0 = \dfrac{10^g}{9}$

3. $1^g = \dfrac{9}{10}^0$

3.4 Circular Measure

1. A circle is simply a plane figure traced by a point that is at a constant distance from a fixed reference point.

2. The fixed reference point is also called the center of the circle.

3. The straight line distance between the center of the circle and the point on the circumference is called the radius of the circle.

4. Any portion of the circumeference of the circle is called an arc.

5. Consider an arc whose length is the same as the radius of the circle. The angle subtended by such an arc at the center of the circle is called a radian. This is represented as 1^c.

    a. The length of the circumference of a circle always bears a constant ratio to its diameter.

    b. Consider two concentric circles, with O being the center.

    c. Let us inscribe two polygons of n-sides, ABCDE.. on the outer circle and abcde.. on the inner circle.

    d. Consider the triangles - ΔOAB and ΔOab.

    e. $\dfrac{AB}{ab} = \dfrac{OA}{oa}$

    f. $\dfrac{\text{perimeter of polygon ABCD..}}{\text{perimeter of polygon abcd..}} = \dfrac{n \times AB}{n \times ab} = \dfrac{OA}{Oa}$

g. As *n* increases indefinitely, we can readily see that perimeter of the polygon tends to be the circumference of the circle.

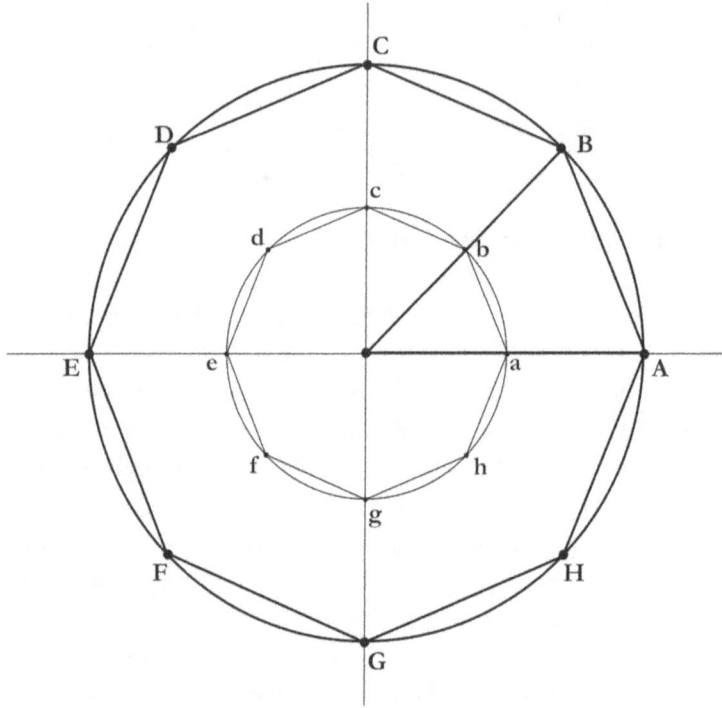

The ratio of the circumference of a circle to its radius is a constant

h. $$\frac{\text{circumference of outer circle}}{\text{circumference of inner circle}} = \frac{n \times AB}{n \times ab}$$

$$= \frac{OA}{Oa} = \frac{\text{radius of outer circle}}{\text{radius of inner circle}}$$

i. $$\frac{\text{circumference of outer circle}}{\text{radius of outer circle}}$$

$$= \frac{\text{circumference of inner circle}}{\text{radius of inner circle}} = \text{constant}$$

j. Therefore, the ratio of the circumference of a circle to its radius is a constant. Since diameter of a circle is twice its radius [constant times the radius]; we can also conclude that the ratio of the circumference of the circle to its diameter is a constant.

k. This constant is called π.

l. The circumference of a circle is always is π times its diameter or $2 \times \pi$ times its radius.

m. The number of radians in an angle is therefore the ratio of the length of the arc which subtends the given angle to its radius.

n. Therefore the circumference of a circle subtends $2 \times \pi$ radians at its center.

3.5 Exercises

1: **Express 75° 15' in right angles**

$$75°15' = 75° + 15 \times \frac{1°}{60} = 75\frac{1°}{4} = \frac{301°}{4}$$

$$= \frac{301°}{4} \times \frac{1}{90} \text{ right angles}$$

$$= \frac{301}{360} \text{ right angle}$$

2: **Express in grades, minutes and seconds: 35° 47' 15"**

$$35°47'15'' = 35° + 47\frac{1'}{4} = 35° + \frac{189'}{4}$$

$$= 35° + \frac{189°}{4 \times 60} = \frac{2863°}{80} \times \frac{100^g}{90}$$

$$= \frac{2863^g}{72} = 39.76388^g = 39^g76'39''$$

3: **Express in right angles and also in minutes and seconds:** $45^g35'24''$

$$45^g35'24'' = 45.3524^g = 45.3524 \times \frac{9°}{10} = 40.8171°$$

$$= 40° + 0.8171 \times 60' = 40°49.296'$$

$$= 40°49' + 0.296 \times 60''$$

$$= 40°49'2''$$

$$45.3524^g = \frac{45.3524}{100} \cdot \text{right angle}$$

$$= 0.453524 \text{ right angle}$$

4:   **Express in radians the angle:** $110^g 30'$

$$110^g 3' = 110.038 = \frac{1103}{10} \times \frac{\pi}{200} \text{ radians}$$

$$= \frac{1103}{2000} \pi \text{ radians}$$

5:   **Which quadrant does the following angle fall in:** 150^g

$$150^g = \frac{150}{100} = 1\frac{1}{2} \text{ right angle}$$

$$2 > 1\frac{1}{2} > 1 \text{ right angles}$$

This is in the second quadrant.

6:   The number of sides in two regular polygons are as 5 : 4, and the difference between their angles is 9°; find the number of sides in the polygons.

Let the number of sides in the polygon be $5x$ and $4x$

Internal angle of polygon with $5x$ sides is:

$$\theta_5 = \frac{\pi(5x-2)}{5x}$$

Internal angle of polygon with $4x$ sides is:

$$\theta_4 = \frac{\pi(4x-2)}{4x}$$

Subtracting:

$$\theta_5 - \theta_4 = 9° = 9 \times \frac{\pi}{180} = \frac{\pi}{20}$$

$$\frac{\pi(5x-2)}{5x} - \frac{\pi(4x-2)}{4x} = \frac{\pi}{20}$$

$$20x - 8 - (20x - 10) = x \Rightarrow x = 2$$

Number of sides in the polygons: 10, 8

4   Basic Trigonometric Ratios

1. A line BY starts from BX and moves in a counter clockwise direction. It creates an angle $\angle YBX$.
2. We drop a perpendicular from a point C on line BY to a point A on the line BX.

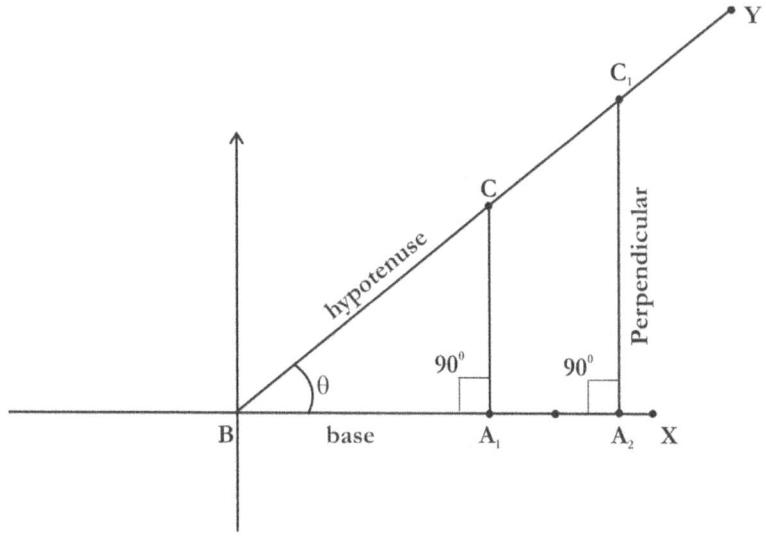

3. The $\triangle ABC$ is a **right-angled triangle**, where BC is the **hypotenuse**, the line CA is called **opposite side** or **the perpendicular** and the line AB is called **adjacent side** or **the base**.
4. The trigonometric ratios of angle $\angle ABC$ are defined as follows:

    a.   Sine of $\angle ABC = \dfrac{\text{perpendicular}}{\text{hypotenuse}}$

    b.   Cosine of $\angle ABC = \dfrac{\text{base}}{\text{hypotenuse}}$

    c.   Tangent of $\angle ABC = \dfrac{\text{perpendicular}}{\text{base}}$

    d.   Secant of $\angle ABC = \dfrac{\text{hypotenuse}}{\text{base}}$

    e.   Cosecant of $\angle ABC = \dfrac{\text{hypotenuse}}{\text{perpendicular}}$

    f.   Cotangent of $\angle ABC = \dfrac{\text{base}}{\text{perpendicular}}$

5.   Therefore, we can make the following observations:

    a.   Cosecant of $\angle ABC = \dfrac{1}{\text{Sine of } \angle ABC}$

    b.   Secant of $\angle ABC = \dfrac{1}{\text{Cosine of } \angle ABC}$

    c.   Cotangent of $\angle ABC = \dfrac{1}{\text{Tangent of } \angle ABC}$

6.   We use a shorthand notation, when we talk of trigonometric ratios. Sine of an angle $\angle ABC$ is simply referred to as $\sin ABC$. Similarly,

    a.   Cosine of $\angle ABC = \cos ABC$

    b.   Tangent of $\angle ABC = \tan ABC$

    c.   Secant of $\angle ABC = \sec ABC$

    d.   Cosecant of $\angle ABC = \csc ABC$

    e.   Cotangent of $\angle ABC = \cot ABC$

7.   The trigonometric ratios are all numbers.

8.   The trigonometric ratios are always the same for a given angle.

9.   Consider the $\triangle ABC$ which we constructed for defining the trigonometric ratios. Let us pick another point C_1 which is on the line BY. Let us drop a perpendicular from C_1 on line BY to a point A_1 on the line BX.

10. Clearly $\triangle ABC$ is similar to $\triangle A_1 BC_1$

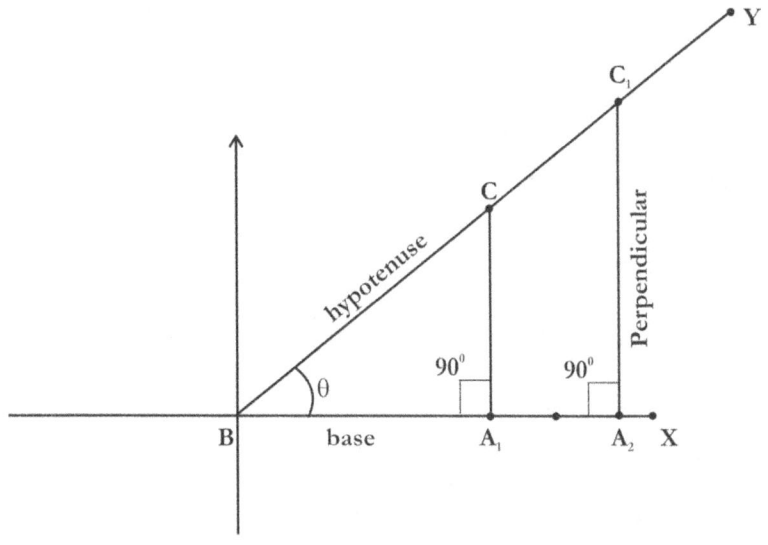

11. $\dfrac{BC}{BC_1} = \dfrac{BA}{BA_1} = \dfrac{CA}{C_1A_1}$

12. The following facts can be readily seen as a consequence of the the previous observation.

    a. $\sin ABC = \dfrac{AC}{BC} = \dfrac{A_1C_1}{BC_1}$

    b. $\cos ABC = \dfrac{AB}{BC} = \dfrac{A_1B}{BC_1}$

    c. $\tan ABC = \dfrac{AC}{AB} = \dfrac{AC_1}{A_1B}$

13. Therefore, we can conclude that the trigonometric ratios are indeed the same for the same angles.

14. Let us now turn our attention to the identify the relationship between the various trigonometric ratios.

15. Let us consider the right angled triangle $\triangle ABC$ that we constructed before.

16. From the pythogorean theorem, we can know that the square on the hypotenuse is equal to the sum of the squares on the other two sides.

17. Therefore, $AB^2 + AC^2 = BC^2$.

18. Dividing both sides of the equation by BC^2, we get:

19. $\dfrac{AB^2}{BC} + \dfrac{AC^2}{BC} = 1$

20. We know that $\dfrac{AB}{BC} = \sin ABC$ and $\dfrac{AC}{BC} = \cos ABC$

21. **Identity 1**: $\sin^2 ABC + \cos^2 ABC = 1$

22. Dividing both sides of the equation by $\cos^2 ABC$, we have:

23. **Identity 2**: $\tan^2 ABC + 1 = \dfrac{1}{\cos^2 ABC} = \sec^2 ABC$

24. Dividing both sides of Identity 1 by $\sin^2 ABC$, we have:

25. **Identity 3**: $1 + \cot^2 ABC = \dfrac{1}{\sin^2 ABC} = \csc^2 ABC$

26. We will look at a few reciprocal identities.

    a. $\csc\theta = \dfrac{1}{\sin\theta}$

    b. $\sin\theta = \dfrac{1}{\csc\theta}$

    c. $\sec\theta = \dfrac{1}{\cos\theta}$

    d. $\cos\theta = \dfrac{1}{\sec\theta}$

    e. $\cot\theta = \dfrac{1}{\tan\theta}$

    f. $\tan\theta = \dfrac{1}{\cot\theta}$

27. Sine Curve: Graphical representation of the limits of sine function

28. Cosine Curve : Graphical representation of the limits of cosine function

29. Tan Curve : Graphical representation of the limits of tangent function

4.1 Expressing Trigonometric ratios - Unit Circle

1. We will start with a unit circle. By unit circle, we mean a circle with a unit radius.

2. Let us consider O to be the center of the circle.

3. From a point p(x,y) – we will call this C, on the circumference, we drop a perpendicular to a point A on the radius---OX along the horizontal. The triangle $\triangle OCA$ is a right angled triangle.

4. The hypotenuse OC has unit length.

5. Let us call angle $\angle COA$ as θ.

6. $\sin\theta = \dfrac{\text{perpendicular}}{\text{hypotenuse}} = \dfrac{\text{perpendicular}}{1} = \text{perpendicular}$.

7. Similarly, $\cos\theta = \dfrac{\text{base}}{\text{hypotenuse}} = \dfrac{\text{base}}{1} = \text{base}$.

8. From pythogorean theorem, we know that $\text{base}^2 + \text{perpendicular}^2 = 1$

9. In the figure below, perpendicular $= x$, and base $= y$

10. $\sin^2\theta + \cos^2\theta = 1$

11. We can make a few observations based on the unit circle example we have seen so far.

12. If we had to determine the least positive angle whose sine is equal to a, then we need to drop a perpendicular whose length is a.

13. If we had to determine the least postive angle whose cosine is equal to a, then we ensure that the base is equal to b.

14. The sine function varies from 0 to 1---the perpendicular can be 0, when points A and C are over each other; and the can be one

when points A and O overlap. The sine function cannot have a value greater than unity.

15. A similar observation can be made about the cosine function as well.

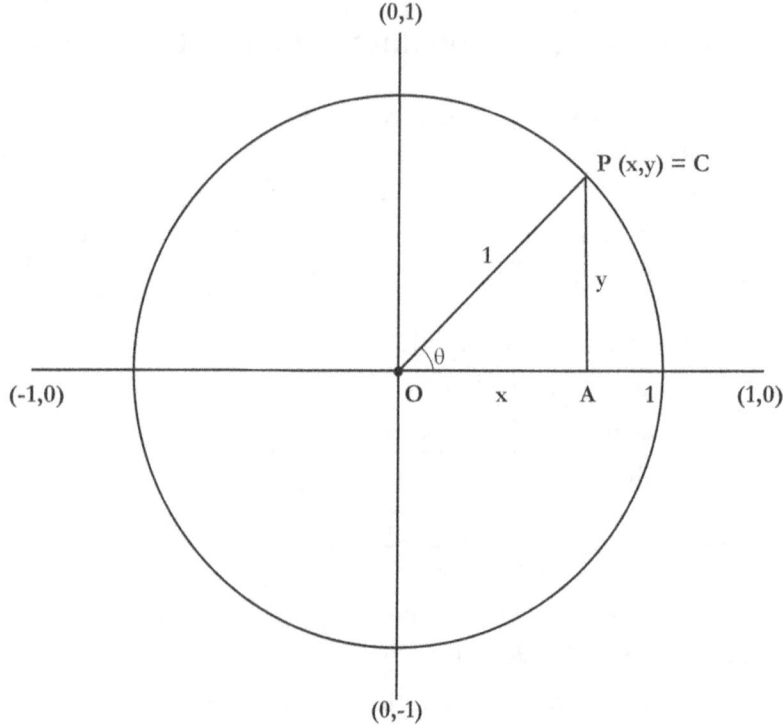

16. Since tangent is the ratio of sine and cosine functions, the tangent is zero, when sine of an angle is zero. Similarly when cosine is zero, the tangent is not defined. Tangent of an angle is equal to one, when sine and cosine are equal to one another. This happens in the case of a right angles isosceles triangle---where the perpendicular is equal to the base of the triangle.

17. It is easy and informative for us to deduce the basic properties of various ratios based on the unit circle.

18. At this point in time, we are considering angles between 0^0 and 90^0 - or angles in the first quadrant only.

19. In a later section, we will deal with angles which are in other quadrants.

4.2 Expressing trigonometric ratios in terms of others

1. For expressing one trigonometric ratio in terms of the others, we start off with the appropriate identity.

2. We will determine the values of sin , cos ratios; all others can be derived using these two basic ratios.

3. For example, let us express all the trigonometric ratios in terms of the sin function.

4. $\sin^2 \theta + \cos^2 \theta = 1$

5. Therefore:

6. $\cos \theta = \sqrt{1 - \sin^2 \theta}$

7. $\tan \theta = \dfrac{\sin \theta}{\cos \theta} = \dfrac{\sin \theta}{\sqrt{1 - \sin^2 \theta}}$

8. The other functions---secant, cosecant and cotangent are reciprocals of cosine, sine and tangent of the respective angles.

9. In the next two sections, we have expressed each of the trigonometric ratio in terms of the others.

10. The student is advised to take a piece of paper and derive each one of these relationship based on his / her understanding. This will also result in gaining an understanding on how to manipulate the trigonometric ratios.

4.3 Trigonometric Ratios in terms of sin, cos and tan

	$\sin\theta$	$\cos\theta$	$\tan\theta$
$\sin\theta$	$\sin\theta$	$\sqrt{1-\cos^2\theta}$	$\dfrac{\tan\theta}{\sqrt{1+\tan^2\theta}}$
$\cos\theta$	$\sqrt{1-\sin^2\theta}$	$\cos\theta$	$\dfrac{1}{1+\tan^2\theta}$
$\tan\theta$	$\dfrac{\sin\theta}{\sqrt{1-\sin^2\theta}}$	$\dfrac{\sqrt{1-\cos^2\theta}}{\cos\theta}$	$\tan\theta$
$\cot\theta$	$\dfrac{\sqrt{1-\sin^2\theta}}{\sin\theta}$	$\dfrac{\cos\theta}{\sqrt{1-\cos^2\theta}}$	$\dfrac{1}{\tan\theta}$
$\sec\theta$	$\dfrac{1}{\sqrt{1-\sin^2\theta}}$	$\dfrac{1}{\cos\theta}$	$\sqrt{1+\tan^2\theta}$
$\csc\theta$	$\dfrac{1}{\sin\theta}$	$\dfrac{1}{\sqrt{1-\cos^2\theta}}$	$\dfrac{\sqrt{1+\tan^2\theta}}{\tan\theta}$

4.4 Trigonometric Ratios in terms of cot, sec and csc

	$\cot\theta$	$\sec\theta$	$\csc\theta$
$\sin\theta$	$\dfrac{1}{\sqrt{1+\cot^2\theta}}$	$\dfrac{\sqrt{\sec^2\theta-1}}{\sec\theta}$	$\dfrac{1}{\csc\theta}$
$\cos\theta$	$\dfrac{\cot\theta}{\sqrt{1+\cot^2\theta}}$	$\dfrac{1}{\sec\theta}$	$\dfrac{\sqrt{\csc^2\theta-1}}{\csc\theta}$
$\tan\theta$	$\dfrac{1}{\cot\theta}$	$\sqrt{\sec^2\theta-1}$	$\dfrac{1}{\sqrt{\csc^2\theta-1}}$
$\cot\theta$	$\cot\theta$	$\dfrac{1}{\sqrt{\sec^2\theta-1}}$	$\sqrt{\csc^2\theta-1}$
$\sec\theta$	$\dfrac{\sqrt{1+\cot^2\theta}}{\cot\theta}$	$\sec\theta$	$\dfrac{\csc\theta}{\sqrt{\csc^2\theta-1}}$
$\csc\theta$	$\sqrt{1+\cot^2\theta}$	$\dfrac{\sqrt{\sec^2\theta-1}}{\sec\theta}$	$\csc\theta$

4.5 Values of trigonometric ratios for a few useful angles

1. Let us consider the triangle $\triangle ABC$ --- we used to define the trigonometric ratios. Let us henceforth refer to $\angle ABC = \theta$.

2. Case 1: $\theta = 45^0$:

    a. We have a right angled isosceles triangle.

    b. In a right angled isosceles triangle, perpendicular:base:hypotenuse $= 1:1:\sqrt{2}$

    c. $\sin\theta = \dfrac{\text{perpendicular}}{\text{hypotenuse}} = \dfrac{1}{\sqrt{2}}$

d. $\cos\theta = \dfrac{\text{base}}{\text{hypotenuse}} = \dfrac{1}{\sqrt{2}}$

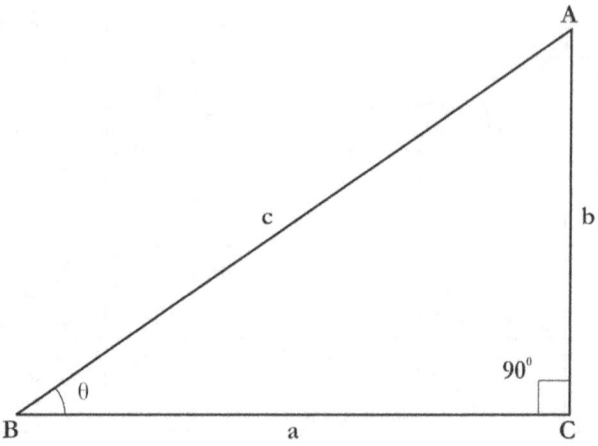

e. $\tan\theta = \dfrac{\sin\theta}{\cos\theta} = 1$

f. $\sec\theta = \dfrac{1}{\cos\theta} = \sqrt{2}$

g. $\csc\theta = \dfrac{1}{\sin\theta} = \sqrt{2}$

h. $\cot\theta = \dfrac{1}{\tan\theta} = 1$

3. Case 2: $\theta = 30^{0}$:

    a. The angles are 30^{0}, 60^{0} and 90^{0}.

    b. In order to find the relationship between various sides, we will do the following. We will extend the perpendicular CA to a point C_1 on the other side of the base. This will mean that $\angle CBA = \angle C_1 BA = 30^{0}$.

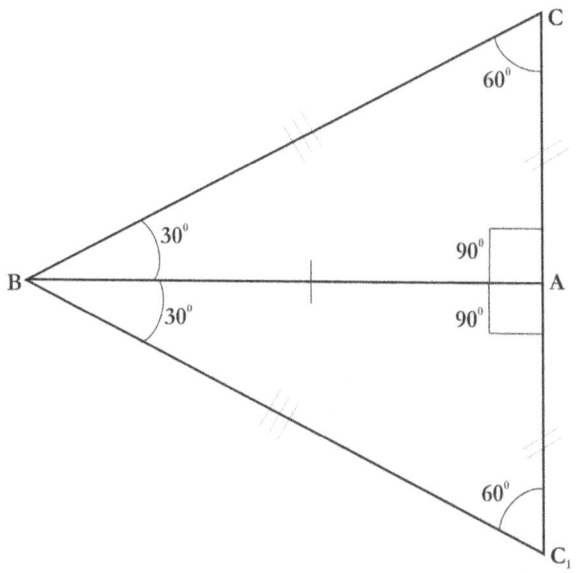

c. $\triangle CC_1B$ is an equilateral triangle. Therefore, the length of the perpendicular AC is half of the hypotenuse BC. Therefore, from the pythogorean theorem, the

$$\text{base} = \sqrt{\text{hypotenuse}^2 - \text{perpendicular}^2}$$
$$= \sqrt{3} \times \text{hypotenuse}$$

d. Therefore, the sides are in the ratio of perpendicular:base:hypotenuse $= 1 : \sqrt{3} : 2$

e. $\sin\theta = \dfrac{\text{perpendicular}}{\text{hypotenuse}} = \dfrac{1}{2}$

f. $\cos\theta = \dfrac{\text{base}}{\text{hypotenuse}} = \dfrac{\sqrt{3}}{\sqrt{2}}$

g. $\tan\theta = \dfrac{\sin\theta}{\cos\theta} = \dfrac{1}{\sqrt{3}}$

h. $\sec\theta = \dfrac{1}{\cos\theta} = \dfrac{2}{\sqrt{3}}$

i. $\csc\theta = \dfrac{1}{\sin\theta} = 2$

j. $\cot\theta = \dfrac{1}{\tan\theta} = \sqrt{3}$

4. Case 3: $\theta = 60^0$:

   a. The angles are 60^0, 30^0 and 90^0

   b. The triangle is the same as the one we considered in case 2. We are now considering the other angle as θ. Therefore the base and perpendicular swap places.

   c. Therefore, the sides are in the ratio of perpendicular:base:hypotenuse $= \sqrt{3} : 1 : 2$

   d. $\sin\theta = \dfrac{\text{perpendicular}}{\text{hypotenuse}} = \dfrac{\sqrt{3}}{2}$

   e. $\cos\theta = \dfrac{\text{base}}{\text{hypotenuse}} = \dfrac{1}{2}$

   f. $\tan\theta = \dfrac{\sin\theta}{\cos\theta} = \sqrt{3}$

   g. $\sec\theta = \dfrac{1}{\cos\theta} = 2$

   h. $\csc\theta = \dfrac{1}{\sin\theta} = \dfrac{2}{\sqrt{3}}$

   i. $\cot\theta = \dfrac{1}{\tan\theta} = \dfrac{1}{\sqrt{3}}$

5. Case 3: $\theta = 0^0$:

   a. When the angle is 0^0, then the perpendicular is 0 and the base is equal to hypotenuse.

   b. $\sin\theta = \dfrac{\text{perpendicular}}{\text{hypotenuse}} = 0$

   c. $\cos\theta = \dfrac{\text{base}}{\text{hypotenuse}} = 1$

   d. $\tan\theta = \dfrac{\sin\theta}{\cos\theta} = 0$

   e. $\sec\theta = \dfrac{1}{\cos\theta} = 1$

   f. $\csc\theta = \dfrac{1}{\sin\theta} = Undefined$

   g. $\cot\theta = \dfrac{1}{\tan\theta} = Undefined$

6.   Case 4: $\theta = 90^{0}$:

   a.   When $\theta = 90^{0}$, the base is zero and the perpendicular is the same as the hypotenuse.

   b. $\sin\theta = \dfrac{\text{perpendicular}}{\text{hypotenuse}} = 1$

   c. $\cos\theta = \dfrac{\text{base}}{\text{hypotenuse}} = 0$

   d. $\tan\theta = \dfrac{\sin\theta}{\cos\theta} = Undefined$

   e. $\sec\theta = \dfrac{1}{\cos\theta} = Undefined$

   f. $\csc\theta = \dfrac{1}{\sin\theta} = 1$

   g. $\cot\theta = \dfrac{1}{\tan\theta} = 0$

4.6 Trigonometric Ratios of Complementary angles

1. When the sum of two angles is equal to 90^0, the two angles are said to be complementary to each other. In the case of a right angled triangle, one of the angles is 90^0. Therefore the other two angles are always complementary.

2. We will now start looking at trigonometric ratios of complementary angles. Angles θ and $90-\theta$ are complementary angles.

3. In a right angled triangle $\triangle ABC$, BC is the hypotenuse.

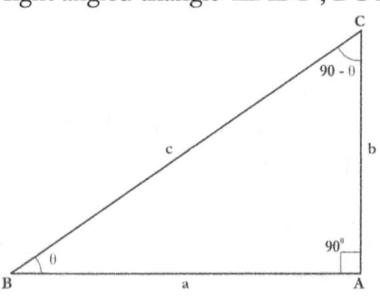

4. Let $\angle ABC = \theta$, then $\angle ACB = 90-\theta$

   a. $\sin\theta = \dfrac{AC}{BC} = \cos 90 - \theta$

   b. $\sin 90 - \theta = \dfrac{AB}{BC} = \cos\theta$

   c. $\tan\theta = \dfrac{AC}{AB} = \cot 90 - \theta$

   d. $\tan 90 - \theta = \dfrac{AB}{AC} = \cot\theta$

   e. From the above relationships, we can deduce:

   f. $\sec\theta = \csc 90 - \theta$

   g. $\csc\theta = \sec 90 - \theta$

5. We can now summarize the findings as follows:

a. The sine of any angle is equal to the cosine of its complement.

b. The tangent of any angle is equal to the cotangent of its complement.

c. The secant of any angle is equal to the cosecant of its complement.

6. The following table captures common trigonometric ratios that we often come across while solving problems. With practice, you will be able to recall these values without much effort.

Degrees	0	30	45	60	90
Radians	0	$\dfrac{\pi}{6}$	$\dfrac{\pi}{4}$	$\dfrac{\pi}{3}$	$\dfrac{\pi}{2}$
sin	0	$\dfrac{1}{2}$	$\dfrac{1}{\sqrt{2}}$	$\dfrac{\sqrt{3}}{2}$	1
cos	1	$\dfrac{\sqrt{3}}{2}$	$\dfrac{1}{\sqrt{2}}$	$\dfrac{1}{2}$	0
tan	0	$\dfrac{1}{\sqrt{3}}$	1	$\sqrt{3}$	∞
cot	∞	$\sqrt{3}$	1	$\dfrac{1}{\sqrt{3}}$	0
csc	∞	2	$\dfrac{1}{\sqrt{2}}$	$\dfrac{2}{\sqrt{3}}$	1
sec	1	$\dfrac{2}{\sqrt{3}}$	$\sqrt{2}$	2	∞

4.7 Exercises

Prove that:

7: $\dfrac{\operatorname{cosec} A}{\cot A + \tan A} = \cos A$

$$\text{LHS} = \frac{\operatorname{cosec} A}{\cot A + \tan A} = \frac{\dfrac{1}{\sin A}}{\dfrac{\cos A}{\sin A} + \dfrac{\sin A}{\cos A}}$$

$$= \frac{\dfrac{1}{\sin A}}{\dfrac{\cos^2 A + \sin^2 A}{\sin A \cos A}} = \frac{1}{\dfrac{1}{\cos A}} = \cos A = \text{RHS}$$

8: $\sec^4 A - \sec^2 A = \tan^4 A + \tan^2 A$

$$\text{LHS} = \sec^4 A - \sec^4 A - \sec^2 A = \sec^2 A (\sec^2 A - 1)$$
$$= (1 + \tan^2 A) \tan^2 A$$
$$= \tan^4 A + \tan^2 A$$
$$= \text{RHS}$$

9: $\dfrac{1}{\operatorname{cosec} A - \cot A} - \dfrac{1}{\sin A} = \dfrac{1}{\sin A} = \dfrac{1}{\operatorname{cosec} A + \cot A}$

$$\text{LHS} = \frac{1}{\operatorname{cosec} A - \cot A} - \frac{1}{\sin A}$$

$$= \frac{\operatorname{cosec} A + \cot A}{\operatorname{cosec} A + \cot A} \times \frac{1}{\operatorname{cosec} A - \cot A} - \operatorname{cosec} A$$

$$= \frac{\operatorname{cosec} A + \cot A}{\operatorname{cosec}^2 A - \cot^2 A} - \operatorname{cosec} A$$

$$= \operatorname{cosec} A + \cot A - \operatorname{cosec} A = \cot A \qquad (1)$$

$$\text{RHS} = \frac{1}{\sin A} - \frac{1}{\operatorname{cosec} A + \cot A}$$

$$= \operatorname{cosec} A - \frac{\operatorname{cosec} A - \cot A}{\operatorname{cosec}^2 A - \cot^2 A}$$

$$= \operatorname{cosec} A - \operatorname{cosec} A + \cot A$$

$$= \cot A \tag{2}$$

LHS = RHS

10: **Express all the ratios in terms of the tangent.**

$$\sec^2 \theta = 1 + \tan^2 \theta$$

$$\sec \theta = \sqrt{1 + \tan^2 \theta}$$

$$\cos \theta = \frac{1}{\sqrt{1 + \tan^2 \theta}}$$

$$\cot \theta = \frac{1}{\tan \theta}$$

$$\sin \theta = \sqrt{1 - \cos^2 \theta} = \sqrt{1 - \frac{1}{1 + \tan^2 \theta}} = \sqrt{\frac{\tan^2 \theta}{1 + \tan^2 \theta}}$$

$$= \frac{\tan \theta}{\sqrt{1 + \tan^2 \theta}}$$

$$\operatorname{cosec} \theta = \frac{\sqrt{1 + \tan^2 \theta}}{\tan \theta}$$

11: **If** $\tan \theta = \dfrac{1}{\sqrt{7}}$ **find the value of** $\dfrac{\operatorname{cosec}^2 \theta - \sec^2 \theta}{\operatorname{cosec}^2 \theta + \sec^2 \theta}$

$$\tan^2 \theta = \frac{1}{7} \quad \cot^2 \theta = 7$$

$$\frac{\operatorname{cosec}^2 \theta - \sec^2 \theta}{\operatorname{cosec}^2 \theta + \sec^2 \theta} = \frac{(1 + \cot^2 \theta) - (1 + \tan^2 \theta)}{(1 + \cot^2 \theta) + (1 + \tan^2 \theta)}$$

$$= \frac{\cot^2 \theta - \tan^2 \theta}{2 + \tan^2 \theta + \cot^2 \theta}$$

$$= \frac{7 - \dfrac{1}{7}}{2 + \dfrac{1}{7} + 7} = \frac{\dfrac{48}{7}}{\dfrac{64}{7}} = \frac{3}{4}$$

5   Trigonometric Ratios across the 4 quadrants

1. In the cartesian coordinate system, we have two axes, the X and Y axis, which are perpendicular to each other and intersect at the origin O.

2. The convention is that X-axis is horizontal and Y-axis is vertical.

3. X-axis coordinates increase in value from left to right and Y-axis increases from bottom to top.

4. Therefore, the points on the X-axis to the left of the origin; and the points on the Y-axis below the origin are negative; the others are positive.

5. Such an intersection, produces 4 quadrants of interest.

    a. First Quadrant : This is the region to the right of the Y-axis and above the X-axis

    b. Second Quadrant : This is the region to the left of the Y-axis axis and above the X-axis

    c. Third Quadrant : This is the region to the left of the Y-axis and below the X-axis

    d. Fourth Quadrant : This is the region to the right of the Y-axis and below the X-axis

6. Let us start with a line OA; where O is the origin of the coordinate system.

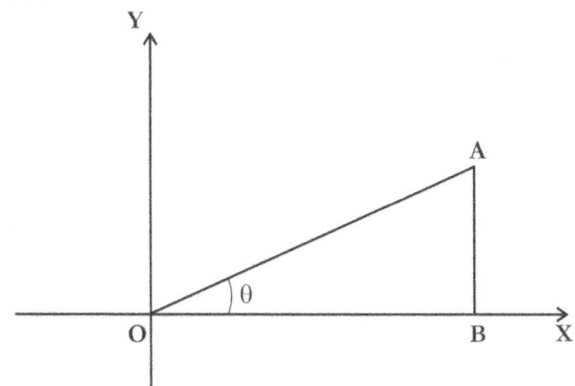

7. Now, let us assume that the line gradually moves in the counter clockwise direction sweeping an angle $\angle AOX$ with the X-axis at all times.

8. Clearly, the point A traces a circle around O.

9. When A is in first quadrant, the base and the perpendicular are positive. Therefore, all trigonometric ratios are positive.

10. When point A is in second quadrant, we find the base to be negative and perpendicular to be positive. Therefore, all the trigonometric ratios, except sine and cosecant, are negative. The sines and cosecants are positive.

11. When the point now traverses to the third quadrant, both--- perpendicular and base are negative. Therefore, tangents and cotangents are positive. Other ratios are negative.

12. When the point moves into the fourth quadrant, we see that the base is positive and perpendicular is negative. The cosines and secants are positive; other trigonometric ratios are negative.

13. Let us now tabulate these findings. This will be a good way to look at the emerging patterns.

Quadrant	I	II	III	IV
sine	0 to 1	1 to 0	0 to -1	-1 to 0
cosine	1 to 0	0 to -1	-1 to 0	0 to 1
tangent	0 to ∞	$-\infty$ to 0	0 to ∞	$-\infty$ to 0

14. One can easily extend this table to include the secant, cosecant and cotangent ratios as well.

Quadrant	I	II	III	IV
secant	1 to ∞	$-\infty$ to -1	-1 to $-\infty$	∞ to 1
cosecant	∞ to 1	1 to ∞	$-\infty$ to-1	-1 to $-\infty$
cotangent	∞ to 0	0 to $-\infty$	∞ to 0	0 to ∞

15. This table also gives us a complete idea of range of values each of the trigonometric ratios can take.

5.1 Exercises

12: **In a triangle one angle contains as many grades as another contains degrees, and the third contains as many centesimal seconds as there are Sexagesimal seconds in the sum of the other two; find the number of radians in each angle.**

Let one angle be in degrees. Expressing it in sexagesimal seconds and radians:

$$\theta_1 = x^\circ = 3600x'' = \frac{\pi x}{180}$$

Let the second angle be in grades. Expressing it in sexagesimal seconds and radians:

$$\theta_2 = x^g = \frac{9}{10}x^\circ = \frac{3600 \times 9x''}{10} = 3240x'' = \frac{\pi x}{200}$$

Adding these two angles:, in sexagesimal seconds

$$\theta_3 = \theta_1 + \theta_2 = 3600x'' + 3240x'' = 6840x''$$

Adding the three angles:

$$\theta_1 + \theta_2 + \theta_3 = 180^\circ$$

$$3600x'' + 3240x'' + 6840x'' = 180 \times 3600''$$

$$x = \frac{1800}{38}$$

Hence, converting to radians:

$$\theta_1 = \frac{1800}{38} \times \frac{\pi}{200} = \frac{9\pi}{38}$$

$$\theta_2 = \frac{1800}{38} \times \frac{\pi}{180} = \frac{10\pi}{38}$$

$$\theta_3 = \theta_1 + \theta_2 = \frac{19\pi}{38} = \frac{\pi}{2}$$

13: Find the number of degrees, minutes, and seconds in the angle at the centre of a circle, whose radius is 5 m, which is subtended by an arc of length 6 m.

Angle subtended:

$$\theta = \frac{\text{arc}}{\text{radius}} = \frac{6}{5} \times \frac{180}{\pi} = 68°45'17'' \text{ radians}$$

14: To turn radians into seconds, prove that we must multiply by 206265 nearly, and to turn seconds into radians the multiplier must be .0000048.

$$\pi^c = 180° = 180 \times 60 \times 60'' = 648000'' \text{ seconds}$$

$$1^c = \frac{648000''}{\pi} = 206265''$$

$$1'' = \frac{\pi}{648000} = \frac{0.0048}{1000} = 0.0000048 \text{ radians}$$

15: If $\sin\theta$ equals $\dfrac{x^2 - y^2}{x^2 + y^2}$, find the values of $\cos\theta$ and $\cot\theta$

$$\cos\theta = \sqrt{1 - \sin^2\theta} = \sqrt{\frac{(x^2 + y^2)^2 - (x^2 - y^2)^2}{(x^2 + y^2)^2}}$$

$$= \sqrt{\frac{4x^2 y^2}{(x^2 + y^2)^2}} = \frac{2xy}{x^2 + y^2}$$

$$\cot\theta = \frac{\cos\theta}{\sin\theta} = \left(\frac{2xy}{x^2 + y^2}\right) \div \left(\frac{x^2 - y^2}{x^2 + y^2}\right) = \frac{2xy}{x^2 - y^2}$$

16:   If $\sin\theta = \dfrac{m^2 + 2mn}{m^2 + 2mn + 2n^2}$ **prove that** $\tan\theta = \dfrac{m^2 + 2mn}{2mn + 2n^2}$

$$\cos^2\theta = 1 - \sin^2\theta = 1 - \frac{(m^2 + 2mn)^2}{(m^2 + 2mn + 2n^2)^2}$$

$$= \frac{(m^2 + 2mn + 2n^2) - (m^2 + 2mn)^2}{(m^2 + 2mn + 2n^2)^2}$$

$$= \frac{(m^2 + 2mn)^2 + 4n^4 + 2.2n^2(m^2 + 2mn) - (m^2 + 2mn)^2}{(m^2 + 2mn + 2n^2)^2}$$

$$= \frac{4n^4 + 4n^2m^2 + 8mn^3}{(m^2 + 2mn + 2n^2)^2}$$

$$= \frac{4n^2(n^2 + m^2 + 2mn)}{(m^2 + 2mn + 2n^2)^2} = \frac{4n^2(n + m)^2}{(m^2 + 2mn + 2n^2)^2}$$

$$\cos\theta = \frac{2n(n + m)}{m^2 + 2mn + 2n^2}$$

$$\tan\theta = \frac{m^2 + 2mn}{2n(n + m)} = \frac{m^2 + 2mn}{2mn + 2n^2}$$

17:   **If** $\cos\theta - \sin\theta = \sqrt{2}\sin\theta,$ **prove**   **that**
$\cos\theta + \sin\theta = \sqrt{2}\cos\theta$

$$\cos\theta - \sin\theta = \sqrt{2}\sin\theta$$

$$\cos\theta = (1 + \sqrt{2})\sin\theta = \frac{(1 + \sqrt{2})\sin\theta}{\sqrt{2} - 1}$$

$$= \frac{2 - 1}{\sqrt{2} - 1}\sin\theta = \frac{\sin\theta}{(\sqrt{2} - 1)}$$

$$\sin\theta = \sqrt{2}\cos\theta - \cos\theta$$

$$\sin\theta + \cos\theta = \sqrt{2}\cos\theta$$

18: **Prove that** $\csc^6\alpha - \cot^6\alpha = 3\csc^2\alpha\cot^2\alpha + 1$

$$\text{LHS} = (\csc^2\alpha)^3 - (\cot^2\alpha)^3$$
$$= (1+\cot^2\alpha)^3 - \cot^6\alpha$$
$$= 1 + \cot^6\alpha + 3\cot^2\alpha(1+\cot^2\alpha) - \cot^6\alpha$$
$$= 1 + 3\cot^2\alpha\csc^2\alpha = \text{RHS}$$

19: **Express** $2\sec^2 A - \sec^4 A - 2\csc^2 A + \csc^4 A$ **in terms of** $\tan A$

LHS
$$= 2\sec^2 A - \sec^4 A - 2\csc^2 A + \csc^4 A$$
$$= -1 + 2\sec^2 A - \sec^4 A + 1 - 2\csc^2 A + \csc^4 A$$
$$= -1(\sec^2 A + 1 - 2\sec^2 A) + (\csc^4 A + 1 - 2\csc^2 A)$$
$$= -(\sec^2 A - 1)^2 + (\csc^2 A - 1)^2$$
$$= -(\tan^2 A)^2 + (\cot^2 A)^2$$
$$= \cot^4 A - \tan^4 A$$
$$= \frac{1}{\tan^4 A} - \tan^4 A$$

20: **Solve the equation** $3\csc^2\theta = 2\sec\theta$

$$\frac{3}{\sin^2\theta} = \frac{2}{\cos\theta}$$
$$(2\sin^2\theta)^2 = (3\cos\theta)^2$$
$$4\sin^4\theta = 9 - 9\sin^2\theta$$

$$4\sin^4\theta + 9\sin^2\theta - 9 = 0$$

$$\sin^2\theta = \frac{-9+\sqrt{81+144}}{8} = \frac{6}{8} = \frac{3}{4}$$

(Taking positive root since $\sin^2\theta$ is positive)
Hence:

$$\sin\theta = \frac{\sqrt{3}}{2} \Rightarrow \theta = \frac{\pi}{3}$$

21:   A man on a cliff observes a boat at an angle of depression of 30°, which is making for the shore immediately beneath him. Three minutes later the angle of depression of the boat is 60°. How soon will it reach the shore?

Let the man be at point A, the shore beneath him at point B. Initially, the boat is seen at point D, later at point C.

$$\tan 30° = \frac{AB}{BD} = \frac{AB}{BC+CD} = \frac{1}{\sqrt{3}}$$

$$\tan 60 = \frac{AB}{BC} = \sqrt{3}$$

$$AB = \sqrt{3}BC$$

$$\sqrt{3}\,BC = \frac{BC+CD}{\sqrt{3}}$$

$$BC = \frac{DC}{2} \Rightarrow DC = 2BC$$

Boat travels the distance CD in 3 minutes.

Boat travels the distance BC in 1.5 minutes (half the time).

22:   **Prove that the equation** $\sin\theta = x + \dfrac{1}{x}$ **is impossible if** x **be real.**

$$\sin\theta = x + \frac{1}{x}$$

$$\sin^2\theta = \left(x + \frac{1}{x}\right)^2 = x^2 + \frac{1}{x^2} + 2$$

For real x, the expression in x is always greater than 2.

$$\sin^2\theta > 2$$

Which is impossible since:

$$\sin^2\theta \leq 1$$

23:   **Show that the equation** $\sec^2\theta = \dfrac{4xy}{(x+y)^2}$ **is only possible when** $x = y$.

$$\sec^2\theta = \frac{4xy}{(x+y)^2}$$

$$\cos^2\theta = \frac{4xy}{(x+y)^2}$$

$$\cos^2\theta = \frac{(x+y)^2}{4xy}$$

$$0 \leq \cos^2\theta \leq 1$$

$$(x+y)^2 \leq 4xy$$

$$(x-y)^2 \leq 0$$

$(x-y)^2$ is a square, hence ≥ 0

Hence it can only be equal to 0.

$$x = y$$

6 Trigonometric functions of angles of any size and sign

1. We can start off with a point A in the first quadrant, from which we will drop a perpendicular AB to the X-axis; the line AO to the origin O is the hypotenuse and OB is the base of the right angled triangle $\triangle AOB$.

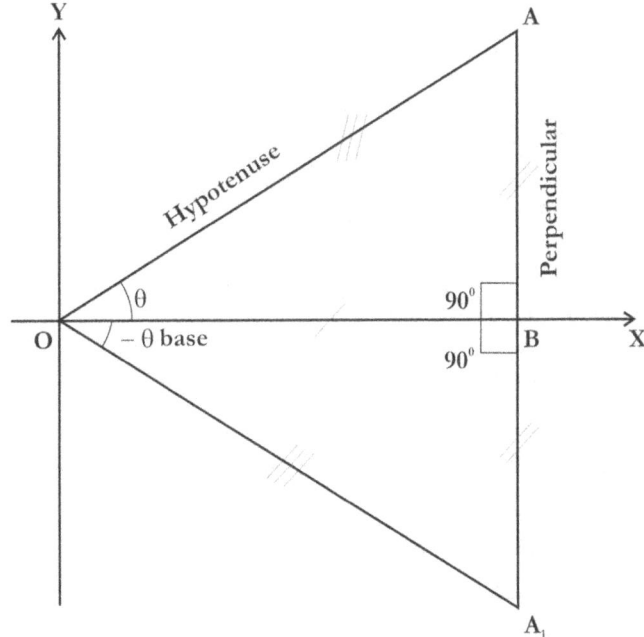

2. Let us plot a point A_1 below the X-axis which is a mirror image of point A. This will mean, $AB = -A_1B$, $AO = A_1 0$ and $OB = OB$.

3. If $\theta = \angle AOB$, then $\angle A_1OB = -\theta$.

   a. $\sin -\theta = \dfrac{A_1B}{OA_1} = \dfrac{-AB}{OA} = -\sin \theta$

b. $\cos-\theta = \dfrac{OB}{OA_1} = \dfrac{OB}{OA} = \cos\theta$

c. $\tan-\theta = \dfrac{A_1B}{OB} = \dfrac{-AB}{OB} = -\tan\theta$

d. $\sec-\theta = \dfrac{OB}{OA_1} = \dfrac{OA}{OA} = \sec\theta$

e. $\csc-\theta = \dfrac{OA_1}{A_1B} = \dfrac{OA}{-AB} = -\csc\theta$

f. $\cot-\theta = \dfrac{OB}{A_1B} = \dfrac{OB}{-AB} = -\cot\theta$

4. If $\theta = \angle AOB$, then $\angle OAB = 90^{0} - \theta$.

a. $\sin 90-\theta = \dfrac{OB}{OA} = \cos\theta$

b. $\cos 90-\theta = \dfrac{AB}{OA} = \sin\theta$

c. $\tan 90-\theta = \dfrac{OB}{AB} = \cot\theta$

d. $\sec 90-\theta = \dfrac{OA}{AB} = \csc\theta$

e. $\csc 90-\theta = \dfrac{OA}{OB} = \sec\theta$

f. $\cot 90-\theta = \dfrac{AB}{OB} = \tan\theta$

5. If $90+\theta$, then .

a. Let us start off with the right angled triangle $\triangle AOB$, where OA is the hypotenuse and $\angle AOB = \theta$.

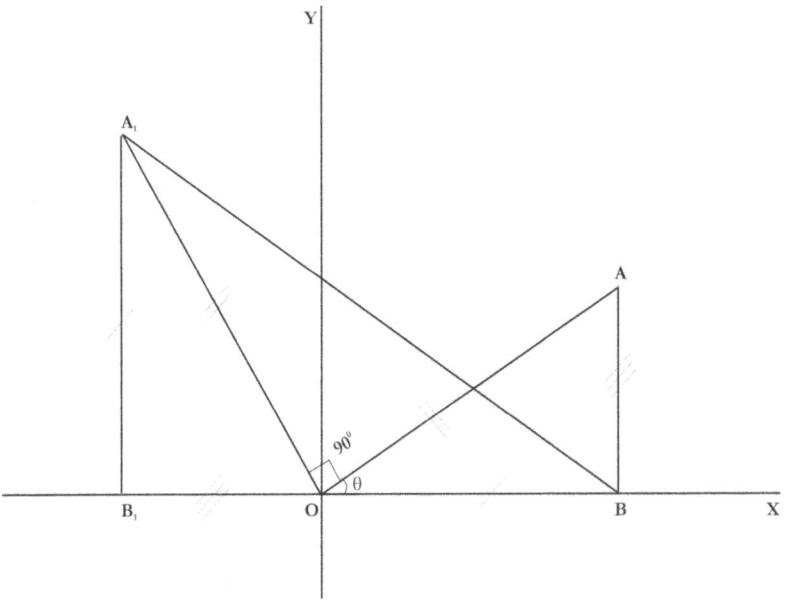

b. Let the point A now move counter clockwise to a position A_1, such that $\angle A_1OB = 90^0 + \theta$.

c. Now drop a perpendicular A_1B_1 to the X-axis.

d. $\triangle AOB$ is similar to $\triangle A_1OB_1$.

e. $AO = A_1O$, $AB = -B_1O$ and $OB = A_1B_1$

f. $\sin(90^0 + \theta) = \dfrac{A_1B_1}{OA_1} = \dfrac{OB}{OA} = \cos\theta$

g. $\cos(90^0 + \theta) = \dfrac{OB_1}{OA_1} = \dfrac{-AB}{OA} = -\sin\theta$

h. $\tan(90^0 + \theta) = \dfrac{A_1B_1}{OB_1} = \dfrac{OB}{-AB} = -\cot\theta$

i. $\sec(90^0 + \theta) = \dfrac{OA_1}{OB_1} = \dfrac{OA}{-AB} = -\csc\theta$

j. $\csc(90^0 + \theta) = \dfrac{OA_1}{A_1B_1} = \dfrac{OA}{OB} = \sec\theta$

$$\text{k.} \quad \cot(90^{\circ}+\theta) = \frac{OB_1}{A_1B_1} = \frac{-AB}{OB} = -\tan\theta$$

6.1 Trigonometric Ratios of supplementary angles

1. When sum of two angles is 180-degrees, then the angles are supplementary.

2. We can look at $180^{\circ}-\theta = 90^{\circ}+(90^{\circ}-\theta)$.

3. Let us $\alpha = (90^{\circ}-\theta)$, then $180^{\circ}-\theta = (90^{\circ}+\alpha)$.

4. We know the formulas for handling $(90^{\circ}+\theta)$ and $(90^{\circ}-\theta)$.

5. Therefore, we can write the trigonometric ratios for the range of angles.

    a. $\sin(180^{\circ}-\theta) = \sin(90^{\circ}+\alpha) = \cos\alpha = \cos(90^{\circ}-\theta) = \sin\theta$

    b. $\cos(180^{\circ}-\theta) = \cos(90^{\circ}+\alpha) = -\sin\alpha$
          $= -\sin(90^{\circ}-\theta) = -\cos\theta$

    c. $\tan(180^{\circ}-\theta) = \tan(90^{\circ}+\alpha) = -\cot\alpha$
          $= -\cot(90^{\circ}-\theta) = -\tan\theta$

    d. $\sec(180^{\circ}-\theta) = \sec(90^{\circ}+\alpha) = -\csc\alpha$
          $= -\csc(90^{\circ}-\theta) = -\sec\theta$

    e. $\csc(180^{\circ}-\theta) = \csc(90^{\circ}+\alpha) = \sec\alpha = \sec(90^{\circ}-\theta) = \csc\theta$

    f. $\cot(180^{\circ}-\theta) = \cot(90^{\circ}+\alpha) = -\tan\alpha$
          $= -\tan(90^{\circ}-\theta) = -\cot\theta$

6. We can look at $180^{\circ}+\theta = 90^{\circ}+(90^{\circ}+\theta)$.

7. By assigning $\alpha = (90^{\circ}+\theta)$, then $180^{\circ}+\theta = (90^{\circ}+\alpha)$

8. We know the formulas for handling $(90^{\circ}+\theta)$ and $(90^{\circ}-\theta)$.

9. Therefore, we can write the trigonometric ratios for the range of angles.

   a. $\sin(180^0 - \theta) = \sin(90^0 + \alpha) = \cos\alpha$
   $= \cos(90^0 + \theta) = -\sin\theta$

   b. $\cos(180^0 - \theta) = \cos(90^0 + \alpha) = -\sin\alpha$
   $= -\sin(90^0 + \theta) = -\cos\theta$

   c. $\tan(180^0 - \theta) = \tan(90^0 + \alpha) = -\cot\alpha$
   $= -\cot(90^0 + \theta) = \tan\theta$

   d. $\sec(180^0 - \theta) = \sec(90^0 + \alpha) = -\csc\alpha$
   $= -\csc(90^0 + \theta) = \sec\theta$

   e. $\csc(180^0 - \theta) = \csc(90^0 + \alpha) = \sec\alpha$
   $= \sec(90^0 + \theta) = -\csc\theta$

   f. $\cot(180^0 - \theta) = \cot(90^0 + \alpha) = -\tan\alpha$
   $= -\tan(90^0 + \theta) = \cot\theta$

10. When the angles are of the form $360 + \theta$, we can simplify the problem using the following observations.

11. Angle around a point is 360^0. So, when a line sweeps an angle $360^0 + \theta$, it makes an angle of θ. Therefore trigonometric ratios $360^0 + \theta$ is the same as trigonometric ratios for θ

    a. $\sin(360^0 + \theta) = \sin\theta$

    b. $\cos(360^0 + \theta) = \cos\theta$

    c. $\tan(360^0 + \theta) = \tan\theta$

    d. $\sec(360^0 + \theta) = \sec\theta$

    e. $\csc(360^0 + \theta) = \csc\theta$

    f. $\cot(360^0 + \theta) = \cot\theta$

12. We can extend this logic further. If the lines went around the point n times and stopped at an angle θ with X-axis, then the total

angle swept is $n \times 360^{0} + \theta$. The trigonometric ratios will be the same as they for the angle θ.

    a. $\sin(n \times 360^{0} + \theta) = \sin \theta$

    b. $\cos(n \times 360^{0} + \theta) = \cos \theta$

    c. $\tan(n \times 360^{0} + \theta) = \tan \theta$

    d. $\sec(n \times 360^{0} + \theta) = \sec \theta$

    e. $\csc(n \times 360^{0} + \theta) = \csc \theta$

    f. $\cot(n \times 360^{0} + \theta) = \cot \theta$

13. Similarly $360 - \theta$ would yield the same ratios as $-\theta$ would.

    a. $\sin(360^{0} - \theta) = -\sin \theta$

    b. $\cos(360^{0} - \theta) = \cos \theta$

    c. $\tan(360^{0} - \theta) = -\tan \theta$

    d. $\sec(360^{0} - \theta) = \sec \theta$

    e. $\csc(360^{0} - \theta) = -\csc \theta$

    f. $\cot(360^{0} - \theta) = -\cot \theta$

14. And as we have seen before, we can extend this logic further. If the lines went around the point n times and stopped at an angle $-\theta$ with X-axis, then the total angle swept is $n \times 360^{0} - \theta$. The trigonometric ratios will be the same as they for the angle $-\theta$.

    a. $\sin(n \times 360^{0} - \theta) = -\sin \theta$

    b. $\cos(n \times 360^{0} - \theta) = \cos \theta$

    c. $\tan(n \times 360^{0} - \theta) = -\tan \theta$

    d. $\sec(n \times 360^{0} - \theta) = \sec \theta$

    e. $\csc(n \times 360^{0} - \theta) = -\csc \theta$

    f. $\cot(n \times 360^{0} - \theta) = -\cot \theta$

15. Let us tabulate these findings. This will serve as a good way to remember the ratios.

	$-\theta$	$\frac{\pi}{2}-\theta$	$\frac{\pi}{2}+\theta$	$\pi-\theta$	$\pi+\theta$	$2\pi+\theta$
$\sin\theta$	$-\sin\theta$	$\cos\theta$	$\cos\theta$	$\sin\theta$	$-\sin\theta$	$\sin\theta$
$\cos\theta$	$\cos\theta$	$\sin\theta$	$-\sin\theta$	$-\cos\theta$	$-\cos\theta$	$\cos\theta$
$\tan\theta$	$-\tan\theta$	$\cot\theta$	$-\cot\theta$	$-\tan\theta$	$\tan\theta$	$\tan\theta$
$\sec\theta$	$\sec\theta$	$\csc\theta$	$-\csc\theta$	$-\sec\theta$	$-\sec\theta$	$\sec\theta$
$\csc\theta$	$-\csc\theta$	$\sec\theta$	$\sec\theta$	$\csc\theta$	$-\csc\theta$	$\csc\theta$
$\cot\theta$	$-\cot\theta$	$\tan\theta$	$-\tan\theta$	$-\cot\theta$	$\cot\theta$	$\cot\theta$

6.2 Common Trigonometric Ratios

The following table captures common trigonometric ratios that we often come across while solving problems. With practice, you will be able to recall these values without much effort.

Deg	0	30	45	60	90	120	135	150	180
Rad	0	$\frac{\pi}{6}$	$\frac{\pi}{4}$	$\frac{\pi}{3}$	$\frac{\pi}{2}$	$\frac{2\pi}{3}$	$\frac{3\pi}{4}$	$\frac{5\pi}{6}$	2π
sin	0	$\frac{1}{2}$	$\frac{1}{\sqrt{2}}$	$\frac{\sqrt{3}}{2}$	1	$\frac{\sqrt{3}}{2}$	$\frac{1}{\sqrt{2}}$	$\frac{1}{2}$	0
cos	1	$\frac{\sqrt{3}}{2}$	$\frac{1}{\sqrt{2}}$	$\frac{1}{2}$	0	$\frac{-1}{2}$	$\frac{-1}{\sqrt{2}}$	$\frac{-\sqrt{3}}{2}$	-1

tan	0	$\dfrac{1}{\sqrt{3}}$	1	$\sqrt{3}$	∞	$\sqrt{3}$	-1	$\dfrac{-1}{\sqrt{3}}$	0
cot	∞	$\sqrt{3}$	1	$\dfrac{1}{\sqrt{3}}$	0	$\dfrac{-1}{\sqrt{3}}$	-1	$-\sqrt{3}$	∞
csc	∞	2	$\dfrac{1}{\sqrt{2}}$	$\dfrac{2}{\sqrt{3}}$	1	$\dfrac{2}{\sqrt{3}}$	$\sqrt{2}$	2	∞
sec	1	$\dfrac{2}{\sqrt{3}}$	$\sqrt{2}$	2	∞	-2	$-\sqrt{2}$	$\dfrac{2}{\sqrt{3}}$	-1

Let us spend a moment to notice and record a few patterns.

1. The sin and cos rows are similar but in reverse order.
2. The tan and cot are reciprocals. Corresponding entries are reciprocals as well.
3. The sec and csc are simply reciprocals of corresponding values of cos and sin .
4. Let us plot the angles in the quadrants so, we have an image associated with them.

Quadrant II: $\dfrac{\pi}{2}+\theta\,;\pi-\theta$	Quadrant I: $\dfrac{\pi}{2}-\theta\,;2\pi+\theta$ and θ
Quadrant III: $\pi+\theta\,;\dfrac{3\pi}{2}-\theta$	Quadrant IV: $2\pi-\theta\,;\dfrac{3\pi}{2}+\theta$ and $-\theta$

5. Remember the mnemonic **ASTC** to remember the signs of trigonometric ratios in the four quadrants.

a. A: In **quadrant 1, A**ll ratios are positive

b. S: In **quadrant 2, S**ine, cosecant are positive, rest negative

c. T: In **quadrant 3, T**angent and cotangent are positive, rest negative

d. C: In **quadrant 4, C**osine and secant are positive, rest negative.

6. Let us look at the memory guide. This will help us determine the ratios and their sign very quickly.

   a. **If angles are** $90^0 + \theta, 90^0 - \theta, 270^0 + \theta, 270^0 - \theta$, **then the ratios change to their co-ratios. sine becomes cosine, tangent becomes cotangent, secant becomes cosecant and so on.**

   b. **If angles are** $180^0 + \theta, 180^0 - \theta, 360^0 + \theta, 360^0 - \theta$ **, then the ratios remains the same. This means sine remains sine, cosine remains cosine and so on.**

   c. **The sign of the ratios is determined using the ASTC mnemonic.**

6.3 Exercises

24: **Prove that:** $\sin 420° \cos 390° + \cos(-300°)\sin(-330°) = 1$

$\text{LHS} = \sin 420° \cos 390° + \cos(-300°)\sin(-330°)$

$= \sin(360° + 60°)\cos(360° + 30°)$

$\quad + \cos(-360° + 60°)(-360° + 30°)$

$= \sin 60° \cos 30° + \cos 60° \sin 30° = \sin(60° + 30°)$

$= \sin 90° = 1 = \text{RHS}$

25: **What are the values of** $\cos A - \sin A$ **and** $\tan A + \cot A$ **when** $A = \dfrac{2\pi}{3}$

$$\cos\frac{2\pi}{3} - \sin\frac{2\pi}{3} = -\frac{1}{2} - \frac{\sqrt{3}}{2} = -\frac{1+\sqrt{3}}{2}$$

$$\tan\frac{2\pi}{3} - \cot\frac{2\pi}{3} = -\sqrt{3} - \frac{1}{\sqrt{3}} = -\frac{4}{\sqrt{3}}$$

26: **What values between** $0°$ **and** $360°$ **may** A **have when** $\cos A = -\dfrac{1}{2}$

$$-\cos(30) = \cos(90 + 30) = -\frac{1}{2}$$

$$A = 120°, 240°$$

27: **Express in terms of the ratios of a positive angle, which is less than 45°, the quantity:** $\tan 137°$

$$\tan(137°) = \tan(90° + 43°) = -\tan(43°)$$

28:   **What sign has** $\sin A + \cos A$ **, given** $A = 278°$

$\sin(278°) + \cos(278°)$

$= \sin(180° + 98°) + \cos(180 + 98°)$

$= -\sin 98° - \cos 98°$

$= -\sin(90° + 8°) - \cos(90° + 8°)$

$= -\cos 8° + \sin 8° = \sin 8° - \cos 8° < 0$

Hence the expression is negative.

7 General Expressions for all Trigonometric Ratios

1. Let us consider a right angled triangle ΔAOB where OA is the hypotenuse, OB is the base and AB is the perpendicular. O is the origin.
2. Let the initial angle $\angle AOB = \theta$.

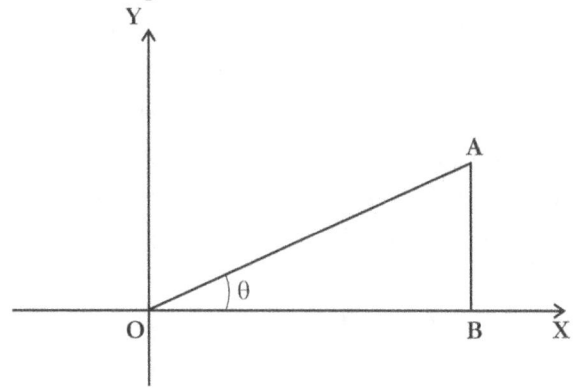

3. Let us now imagine that the line OA starts to rotate in the counter clockwise direction.
4. This revolving line coincides with the original position after $2 \times \pi$ radians after n rotations about point O.
5. The generalized notation for angle θ is $2n\pi + \theta$, where n is an integer. If n is negative, it means that the line has been rotating in the clockwise direction.
6. Equipped with this mental model, it is time for us to proceed to the next steps in making use of this fact in the context of determining the trigonometric ratios.

General Expressions for angles that have the same Sine

1. Case 1: When we have angle that is of the form $n \times (2 \times \pi) + \theta$, this reduces to the an angle that is $360^0 + \theta$. We have see the formulas for this in the previous sections.

    a. $\sin(2n\pi + \theta) = \sin(\theta)$

    b. $\sin(2n\pi - \theta) = -\sin(\theta)$

2.   Case 2: When we have an angle that is of the form $2 \times (n+1) \times \pi + \theta)$, we can see that this reduces to an angle of $180^0 + \theta$. We have see the formulas for this in the previous sections.

    a. $\sin((2n+1)\pi + \theta) = -\sin(\theta)$

    b. $\sin((2n+1)\pi - \theta) = \sin(\theta)$

3.   Clearly, Case 1-i and case 2-ii produces the same value of sine.

4.   Merging these two cases, we can deduce that the general expression for the angles that have the same sine is: $n\pi + (-1)^n \theta$

General Expressions for angles that have the same Cosine

1.   Case 1: When we have angle that is of the form $n \times (2 \times \pi) + \theta$, this reduces to the an angle that is $360^0 + \theta$. We have see the formulas for this in the previous sections.

    a. $\cos(2n\pi + \theta) = \cos(\theta)$

    b. $\cos(2n\pi - \theta) = \cos(\theta)$

2.   Case 2: When we have an angle that is of the form $2 \times (n+1) \times \pi + \theta)$, we can see that this reduces to an angle of $180^0 + \theta$. We have see the formulas for this in the previous sections.

    a. $\cos((2n+1)\pi + \theta) = -\cos(\theta)$

    b. $\cos((2n+1)\pi - \theta) = -\cos(\theta)$

3.   iii. Clearly, here cosine function changes sign. So, these angles do not produce the same cosine.

4.   Clearly Case 1 produces the same value of cosine.

5.   Merging these two cases, we can deduce that the general expression for the angles that have the same cosine is: $2n\pi \pm \theta$

General Expressions for Tangent

1.   Case 1: When we have angle that is of the form $n \times (2 \times \pi) + \theta$, this reduces to the an angle that is $360^0 + \theta$. We have see the formulas for this in the previous sections.

    a. $\tan(2n\pi + \theta) = \tan(\theta)$

    b. $\tan(2n\pi - \theta) = -\tan(\theta)$

2. Case 2: When we have an angle that is of the form $2 \times (n+1) \times \pi + \theta)$, we can see that this reduces to an angle of $180^0 + \theta$. We have see the formulas for this in the previous sections.

    a. $\tan((2n+1)\pi + \theta) = \tan(\theta)$

    b. $\tan((2n+1)\pi - \theta) = -\tan(\theta)$

3. Clearly, Case 1-i and Case 2-i produces the same tangent.

4. Merging these two cases, we can deduce that the general expression for the angles that have the same sine is: $(2n+1)\pi + \theta$

7.1 Exercises

29:   **Find the general value of θ that satisfies:** $4\sin^2 \theta = 3$

$$\sin\theta = \pm\frac{\sqrt{3}}{2}$$

$$\theta = \pm\frac{\pi}{3}$$

General Value:

$$\theta = n\pi \pm (-1)^n \frac{\pi}{3}$$

30:   **Find the general value of θ that satisfies:** $2\sqrt{3}\cos^2 \theta = \sin\theta$

$$2\sqrt{3}(1 - \sin^2 \theta) = \sin\theta$$

$$2\sqrt{3}\sin^2 + \sin\theta - 2\sqrt{3} = 0$$

$$\sin\theta = \frac{-1\pm\sqrt{1+48}}{4\sqrt{3}} = \frac{-1\pm7}{4\sqrt{3}} = -\frac{2}{\sqrt{3}}, \frac{\sqrt{3}}{2}$$

Discarding the impossible value, we have:

$$\sin\theta = \frac{\sqrt{3}}{2} \Rightarrow \theta = \frac{\pi}{3}$$

$$\theta = n\pi + (-1)^n \frac{\pi}{3}$$

31: $\cos 5\theta = \cos 4\theta$

$5\theta = 2n\pi \pm 4\theta$

Consider the negative sign:

$$5\theta = 2n\pi - 4\theta \Rightarrow \theta = \frac{2n\pi}{9}$$

Consider the positive sign:

$$5\theta = 2n\pi + 4\theta \Rightarrow \theta = 2n\pi$$

Hence:

$$\theta = \frac{2n\pi}{9}, 2n\pi$$

32:   **Solve the equation:** $\tan mx + \cot nx = 0$

$$\tan mx = -\cot nx = \tan\left(\frac{\pi}{2} + nx\right)$$

$$mx = p\pi + \frac{\pi}{2} + nx$$

$$x = \frac{\pi}{m-n}\left(p + \frac{1}{2}\right) = \frac{(2p+1)\pi}{2(m-n)}$$

33:   **If** $\tan^2\theta = \frac{5}{4}$ **find** $\text{vers}\,\theta$

$$\tan^2\theta = \frac{5}{4}$$

$$\sec^2\theta = 1 + \tan^2\theta = \frac{9}{4}$$

$$\sec\theta = \pm\frac{3}{2} \Rightarrow \cos\theta = \pm\frac{2}{3}$$

$$\text{vers}\,\theta = 1 - \cos\theta = 1\pm\frac{2}{3} = \frac{1}{3},\frac{5}{3}$$

8 Trigonometric Ratio: Sum and difference of two angles

1. For exploring the topic of trignometric ratios of sum and difference of two angles, we must begin with a picture representing the problem space.

2. Let us consider a right angled triangle ΔAOB where OA is the hypotenuse, OB is the base and AB is the perpendicular. O is the origin.

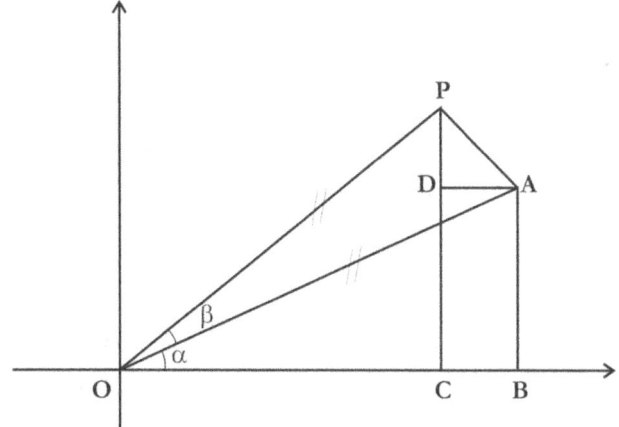

3. Let the initial angle $\angle AOB = \alpha$.

4. Let us now imagine that the line OA starts to rotate in the counter clockwise direction to a point P, such that $\angle POA = \beta$. We drop a perpendicular from P to point C to the X-axis. Therefore, triangle ΔPOC is a right angled triangle and angle $\angle POC = \alpha + \beta$.

5. Draw a line AD parallel to BC. This means that ΔAPD is a right angled triangle.

6. From the figure, we can make the following observations:

   a. $\angle AOB = \alpha$

   b. $\angle POC = \beta$

   c. $\angle OPC = 90^{0} - (\alpha + \beta)$

      d. $\angle DPA = \alpha$

7.   Let us determine the value of $\sin(A+B)$.

8. $\sin(\alpha+\beta) = \dfrac{PC}{OP} = \dfrac{PD+DC}{OP} = \dfrac{PD}{OP} + \dfrac{AB}{OP}, \because DC = AB.$

$$= \dfrac{PD.AP}{AP.OP} + \dfrac{AB.OA}{OA.OP}$$

$$= \dfrac{PD}{AP}\dfrac{AP}{OP} + \dfrac{AB}{OA}\dfrac{OA}{OP}$$

$$= \cos\alpha\sin\beta + \sin\alpha\cos\beta$$

9.   Let   us   determine   the   value   of $\cos(A+B)$.

$$\cos(\alpha+\beta) = \dfrac{OC}{OP} = \dfrac{OB-BC}{OP}$$

$$= \dfrac{OB}{OP} - \dfrac{BC}{OP} = \dfrac{OB.OA}{OA.OP} - \dfrac{AD.AP}{AP.OP}, \because BC = AD$$

$$= \dfrac{OB}{OA}\dfrac{OA}{OP} - \dfrac{AD}{AP}\dfrac{AP}{OP}$$

$$= \cos\alpha\cos\beta - \sin\alpha\sin\beta$$

10. We know, that $\sin(-\beta) = -\sin\beta$ and $\cos(-\beta) = \cos(\beta)$

$$\therefore \sin(\alpha-\beta) = \sin(\alpha+(-\beta)) = \sin\alpha\cos\beta - \cos\alpha\sin\beta$$

$$\cos(\alpha-\beta) = \cos(\alpha+(-\beta)) = \cos\alpha\cos\beta + \sin\alpha\sin\beta$$

11. Summary :

      a. $\sin(\alpha+\beta) = \sin\alpha\cos\beta + \cos\alpha\sin\beta$

      b. $\cos(\alpha+\beta) = \cos\alpha\cos\beta - \sin\alpha\sin\beta$

      c. $\sin(\alpha-\beta) = \sin\alpha\cos\beta - \cos\alpha\sin\beta$

      d. $\cos(\alpha-\beta) = \cos\alpha\cos\beta + \sin\alpha\sin\beta$

12. Let $\alpha = \dfrac{\gamma+\delta}{2}$ and $\beta = \dfrac{\gamma-\delta}{2}$

      $\alpha+\beta = \gamma$ and $\alpha-\beta = \delta$

Substituting these values in the equations for $\sin(\alpha+\beta)$, $\cos(\alpha+\beta)$, $\sin(\alpha-\beta)$ and $\cos(\alpha-\beta)$, we have the following:

$$\sin\gamma + \sin\delta = 2\sin(\frac{\gamma+\delta}{2})\cos(\frac{\gamma-\delta}{2})$$

$$\cos\gamma + \cos\delta = 2\cos(\frac{\gamma+\delta}{2})\cos(\frac{\gamma-\delta}{2})$$

$$\sin\gamma - \sin\delta = 2\cos(\frac{\gamma+\delta}{2})\sin(\frac{\gamma-\delta}{2})$$

$$\cos\gamma - \cos\delta = -2\sin(\frac{\gamma+\delta}{2})\cos(\frac{\gamma-\delta}{2})$$

13. Some books drop the negative sign in the fourth equation, absorb the negative sign into the sine function and represent the same as

$$\cos\gamma - \cos\delta = 2\sin(\frac{\gamma+\delta}{2})\cos(\frac{\delta-\gamma}{2})$$

14. Let us now turn our attention to tangents; and determine the equation for $\tan(\alpha+\beta)$.

$$\tan(\alpha+\beta) = \frac{\sin(\alpha+\beta)}{\cos(\alpha+\beta)}$$

$$= \frac{\sin\alpha\cos\beta + \cos\alpha\sin\beta}{\cos\alpha\cos\beta - \sin\alpha\sin\beta}$$

Divide the numerator and denominator by $\cos\alpha\cos\beta$, we get,

$$= \frac{\dfrac{\sin\alpha}{\cos\alpha} + \dfrac{\sin\beta}{\cos\beta}}{1 - \dfrac{\dfrac{\sin\alpha}{\cos\alpha}}{\dfrac{\sin\beta}{\cos\beta}}}$$

$$= \frac{\tan\alpha + \tan\beta}{1 - \tan\alpha\tan\beta}$$

15. $\tan(\alpha-\beta) = \dfrac{\sin(\alpha-\beta)}{\cos(\alpha-\beta)}$

$$= \frac{\sin\alpha\cos\beta - \cos\alpha\sin\beta}{\cos\alpha\cos\beta + \sin\alpha\sin\beta}$$

Divide the numerator and denominator by $\cos\alpha\cos\beta$, we get,

$$= \frac{\dfrac{\sin\alpha}{\cos\alpha} - \dfrac{\sin\beta}{\cos\beta}}{1 + \dfrac{\dfrac{\sin\alpha}{\cos\alpha}}{\dfrac{\sin\beta}{\cos\beta}}} = \frac{\tan\alpha - \tan\beta}{1 + \tan\alpha\tan\beta}$$

16. Summary:

    a. $\tan(\alpha + \beta) = \dfrac{\tan\alpha + \tan\beta}{1 - \tan\alpha\tan\beta}$

    b. $\tan(\alpha - \beta) = \dfrac{\tan\alpha - \tan\beta}{1 + \tan\alpha\tan\beta}$

17. Let us now consider three angles α, β and γ. We will now determine the trigonometric equations for angles $\alpha + \beta + \gamma$.

$$\sin(\alpha + \beta + \gamma)$$
$$= \sin(\alpha + \beta)\cos\gamma + \cos(\alpha + \beta)\sin\gamma$$
$$= (\sin\alpha\cos\beta + \cos\alpha\sin\beta)\cos\gamma$$
$$\qquad + (\cos\alpha\cos\beta - \sin\alpha\sin\beta)\sin\gamma$$
$$= \sin\alpha\cos\beta\cos\gamma + \cos\alpha\sin\beta\sin\gamma + \cos\alpha\cos\beta\sin\gamma$$
$$\qquad - \sin\alpha\sin\beta\sin\gamma$$

18. $\cos(\alpha + \beta + \gamma)$
$$= \cos(\alpha + \beta)\cos\gamma - \sin(\alpha + \beta)\sin\gamma$$
$$= (\cos\alpha\cos\beta - \sin\alpha\sin\beta)\cos\gamma$$
$$\qquad + (\sin\alpha\cos\beta + \cos\alpha\sin\beta)\sin\gamma$$
$$= \cos\alpha\cos\beta\cos\gamma - \cos\alpha\sin\beta\sin\gamma - \sin\alpha\cos\beta\sin\gamma$$
$$\qquad - \sin\alpha\sin\beta\sin\gamma$$

19. $\tan(\alpha + \beta + \gamma)$

$$= \frac{\tan(\alpha+\beta)+\tan\gamma}{1-\tan(\alpha+\beta)\tan\gamma}$$

$$= \frac{\dfrac{\tan\alpha+\tan\beta}{1-\tan\alpha\tan\beta}+\tan\gamma}{1-\dfrac{\tan\alpha+\tan\beta}{1-\tan\alpha\tan\beta}\tan\gamma}$$

$$= \frac{\tan\alpha+\tan\beta+\tan\gamma}{1-\tan\alpha\tan\beta-\tan\beta\tan\gamma-\tan\gamma\tan\alpha}$$

8.1 Exercises

34: **If** $\sin\alpha=\dfrac{45}{33}$ **and** $\sin\beta=\dfrac{33}{65}$ **, find the value of** $\sin(\alpha-\beta)$ **and** $\sin(\alpha+\beta)$

$$\cos\alpha = \sqrt{1-\frac{45^2}{53^2}} = \frac{28}{53}$$

$$\cos\beta = \sqrt{1-\frac{33^2}{65^2}} = \frac{56}{65}$$

$$\sin(\alpha-\beta) = \sin\alpha\cos\beta - \cos\alpha\sin\beta$$

$$= \frac{45}{53}\cdot\frac{56}{65} - \frac{28}{53}\cdot\frac{33}{65}$$

$$= \frac{1596}{3445}$$

$$\sin(\alpha+\beta) = \sin\alpha\cos\beta + \cos\alpha\sin\beta$$

$$= \frac{45}{53}\cdot\frac{56}{65} + \frac{28}{53}\cdot\frac{33}{65}$$

$$= \frac{3444}{34445}$$

Plane Trigonometry

35: **Prove that:**

$$\sin(n+1)A\sin(n-1)A+\cos(n+1)A\cos(n-1)A=\cos 2A$$

LHS is of the form

$$\cos\alpha\cos\beta+\sin\alpha\sin\beta=\cos(\alpha-\beta)$$

where $\alpha=(n+1)A$ and $\beta=(n-1)A$

$$\text{LHS}=\cos((n+1)A-(n-1)A)$$
$$=\cos(nA+A-nA+A)$$
$$=\cos 2A$$
$$=\text{RHS}$$

36: **Prove that** $\dfrac{\cos 2B+\cos 2A}{\cos 2B-\cos 2A}=\cot(A+B)\cot(A-B)$

$$\text{LHS}=\frac{2\cos\dfrac{2A+2B}{2}\cos\dfrac{2A-2B}{2}}{-2\sin\dfrac{2B+2A}{2}\sin\dfrac{2B-2A}{2}}$$

$$=\frac{\cos(A+B)(\cos A-B)}{-\sin(A+B)\sin(B-A)}$$

$$=\frac{\cos A+B}{\sin A+B}\cdot\frac{\cos(A-B)}{\sin(A-B)}$$

$$=\cot(A+B)\ \cot(A-B)=\text{RHS}$$

37: $\dfrac{\cos A+\cos B}{\cos B-\cos A}=\cot\dfrac{A+B}{2}\cot\dfrac{A-B}{2}$

$$\text{LHS}=\frac{2\cos\dfrac{A+B}{2}\cos\dfrac{A-B}{2}}{2\sin\dfrac{A+B}{2}\sin\dfrac{A-B}{2}}$$

$$= \cot\frac{A+B}{2}\cot\frac{A-B}{2} = \text{RHS}$$

38: $\quad \sin 10° + \sin 20° + \sin 40° + \sin 50° = \sin 70° + \sin 80°$

Regrouping, the LHS

$$= \sin 50° + \sin 10° + \sin 40° + \sin 20°$$
$$= 2\sin 30° \cos 20° + 2\sin 30° \cos 10°$$
$$= \cos 20° + \cos 10°$$
$$= \cos(90° - 70°) + \cos(90° - 80°)$$
$$= \sin 70° + \sin 80° = \text{RHS}$$

Plane Trigonometry

9 Trigonometric Ratio: Multiple and Sub-multiple angles

9.1 Multiple angles

1. The formulas for trigonometric ratios of multiple angles can be deduced easily.
2. If $\alpha = \beta$, then $\alpha + \beta = 2\alpha$
3. Therefore:

   a. $\sin(2\alpha) = \sin(\alpha + \alpha)$
   $$= \sin\alpha\cos\alpha + \sin\alpha\cos\alpha = 2\sin\alpha\cos\alpha$$

   b. $\cos(2\alpha) = \cos(\alpha + \alpha) = \cos\alpha\cos\alpha - \sin\alpha\sin\alpha$
   $$= \cos^2\alpha - \sin^2\alpha = 1 - 2\sin^2\alpha = 2\cos^2\alpha - 1$$

   c. $\tan(2\alpha) = \dfrac{\tan\alpha + \tan\alpha}{1 - \tan\alpha\tan\alpha} = \dfrac{2\tan\alpha}{1 - \tan^2\alpha}$

4. Similarly, we can solve for trigonometric ratios for angles which are $3 \times \alpha$ by representing $3\alpha = (2\alpha + \alpha)$

   d. $\sin(3\alpha) = \sin(\alpha + 2\alpha)$
   $$= \sin\alpha\cos 2\alpha + \cos\alpha\sin 2\alpha$$
   $$= \sin\alpha(1 - 2\sin^2\alpha) + \cos\alpha.2\sin\alpha\cos\alpha$$
   $$= \sin\alpha(1 - 2\sin^2\alpha) + 2\sin\alpha(1 - \sin^2\alpha)$$
   $$= 3\sin\alpha - 4\sin^3\alpha$$

   e. $\cos(3\alpha) = \cos(\alpha + 2\alpha)$
   $$= \cos\alpha\cos 2\alpha - \sin\alpha\sin 2\alpha$$
   $$= \cos\alpha(2\cos^2\alpha - 1) - \sin\alpha.2\sin\alpha\cos\alpha$$
   $$= \cos\alpha(2\cos^2\alpha - 1) - 2\cos\alpha(1 - \cos^2\alpha)$$
   $$= 4\cos^3\alpha - 3\cos\alpha$$

   f. $\tan(3\alpha) = \tan(\alpha + 2\alpha)$

$$= \frac{\tan\alpha + \tan 2\alpha}{1 - \tan\alpha\tan 2\alpha}$$

$$= \frac{\tan\alpha + \dfrac{2\tan\alpha}{1 - \tan^2\alpha}}{1 - \tan\alpha \cdot \dfrac{2\tan\alpha}{1 - \tan^2\alpha}}$$

$$= \frac{\tan\alpha(1 - \tan^2\alpha) + 2\tan\alpha}{1 - \tan^2\alpha - 2\tan^2\alpha}$$

$$= \frac{3\tan\alpha - \tan^3\alpha}{1 - 3\tan^3\alpha}$$

5. Summary of equations:

    g. $\sin(2\alpha) = 2\sin\alpha\cos\alpha$

    h. $\cos(2\alpha) = \cos^2\alpha - \sin^2\alpha$

$$= 1 - 2\sin^2\alpha = 2\cos^2\alpha - 1$$

    i. $\tan(2\alpha) = \dfrac{2\tan\alpha}{1 - \tan^2\alpha}$

    j. $\sin(3\alpha) = 3\sin\alpha - 4\sin^3\alpha$

    k. $\cos(3\alpha) = 4\cos^3\alpha - 3\cos\alpha$

    l. $\tan(3\alpha) = \dfrac{3\tan\alpha - \tan^3\alpha}{1 - 3\tan^3\alpha}$

9.2   Submultiple angles

1. Now, in the aforementioned equations, we substitute $2\alpha = \beta$, then $\alpha = \dfrac{\beta}{2}$ and we can derive the equations for submultiple angles.

2. Therefore, we have the following relationships.

    a. $\sin\alpha = 2\sin\dfrac{\alpha}{2}\cos\dfrac{\alpha}{2}$

    b. $\cos\alpha = \cos^2\dfrac{\alpha}{2} - \sin^2\dfrac{\alpha}{2}$

    $\qquad = 2\cos^2\dfrac{\alpha}{2} - 1$

    $\qquad = 1 - 2\sin^2\dfrac{\alpha}{2}$

    c. $\tan\alpha = \dfrac{2\tan\dfrac{\alpha}{2}}{1 - \tan^2\dfrac{\alpha}{2}}$

3. From the equation, $\cos\alpha = 1 - 2\sin^2\dfrac{\alpha}{2}$. we get

    $$\sin\dfrac{\alpha}{2} = \dfrac{\pm\sqrt{1-\cos\alpha}}{2}$$

    And from the relationship, $\cos\alpha = 2\cos^2\dfrac{\alpha}{2} - 1$, We get

    $$\cos\dfrac{\alpha}{2} = \dfrac{\pm\sqrt{1+\cos\alpha}}{2}$$

4. From the above two equations, we can see that

    $$\tan\dfrac{\alpha}{2} = \dfrac{\sin\dfrac{\alpha}{2}}{\cos\dfrac{\alpha}{2}} = \pm\dfrac{\sqrt{1-\cos\alpha}}{\sqrt{1+\cos\alpha}}$$

5. $\sin \alpha = 2\sin\dfrac{\alpha}{2}\cos\dfrac{\alpha}{2} \ \ldots$ (1)

$\sin^2\dfrac{\alpha}{2} + \cos^2\dfrac{\alpha}{2} = 1 \ \ldots$ (2)

Adding (1) and (2), we get

$$\sin^2\frac{\alpha}{2} + \cos^2\frac{\alpha}{2} + 2\sin\frac{\alpha}{2}\cos\frac{\alpha}{2} = 1 + \sin\alpha$$

$$\left(\sin\frac{\alpha}{2} + \cos\frac{\alpha}{2}\right)^2 = (1 + \sin\alpha)$$

$$\sin\frac{\alpha}{2} + \cos\frac{\alpha}{2} = \pm\sqrt{(1 + \sin\alpha)} \ \ldots \quad (3)$$

Subtracting (1) from (2), we get

$$\sin^2\frac{\alpha}{2} + \cos^2\frac{\alpha}{2} - 2\sin\frac{\alpha}{2}\cos\frac{\alpha}{2} = 1 - \sin\alpha$$

$$\left(\sin\frac{\alpha}{2} - \cos\frac{\alpha}{2}\right)^2 = (1 - \sin\alpha)$$

$$\sin\frac{\alpha}{2} + \cos\frac{\alpha}{2} = \pm\sqrt{(1 - \sin\alpha)} \ \ldots \ (4)$$

Adding (3) and (4), we get

$$2\sin\frac{\alpha}{2} = \pm\sqrt{(1 + \sin\alpha)} + \pm\sqrt{(1 - \sin\alpha)}$$

Subtracting (4) from (3), we get

$$2\cos\frac{\alpha}{2} = \pm\sqrt{(1 + \sin\alpha)} - \pm\sqrt{(1 - \sin\alpha)}$$

9.3 Exercises

39: $\dfrac{\sin 2A}{1+\cos 2A} = \tan A$

$$\text{LHS} = \frac{\sin 2A}{1+\cos 2A} = \frac{2\sin A\cos A}{2\cos^2 A} = \frac{\sin A}{\cos A} = \tan A$$
$$= \text{RHS}$$

40: $\dfrac{\sec 8A-1}{\sec 4A-1} = \dfrac{\tan 8A}{\tan 2A}$

$$\text{LHS} = \frac{\dfrac{1}{\cos 8A}-1}{\dfrac{1}{\cos 4A}-1} = \frac{\cos 4A(1-\cos 8A)}{\cos 8A(1-\cos 4A)}$$

$$\frac{(1-1+2\sin^2 4A)\cos 4A}{1-1+2\sin^2 2A\cdot \cos 8A}$$

$$= \frac{2\sin 4A\cos 4A\sin 4A}{2\sin^2 2A\cos 8A}$$

$$= \frac{\sin 8A}{\cos 8A}\cdot \frac{\sin 4A}{2\sin^2 2A}$$

$$= \tan 8A\cdot \frac{2\sin 2A\cos 2A}{2\sin^2 2A} = \frac{\tan 8A}{\tan 2A} = \text{RHS}$$

41: $2\cos 7\theta \sin 5\theta$

$$= \sin(7\theta + 5\theta) - \sin(7\theta + 5\theta)$$
$$= \sin 12\theta - \sin 2\theta$$

42: If $\cos\alpha = \dfrac{11}{61}$ and $\sin\beta = \dfrac{4}{5}$, find the values of $\sin^2\dfrac{\alpha-\beta}{2}$

and $\cos^2\dfrac{\alpha+\beta}{2}$, the angles α and β being positive acute angles.

$$\cos\alpha = \frac{11}{61}, \sin\beta = \frac{4}{5}$$

$$\sin\alpha = \sqrt{1-\cos^2\alpha} = \sqrt{1-\frac{121}{3721}} = \frac{60}{61}$$

$$\cos\beta = \sqrt{1-\sin^2\beta} = \sqrt{1-\frac{16}{25}} = \frac{3}{5}$$

$$\sin^2\left(\frac{\alpha-\beta}{2}\right) = \frac{1-\cos(\alpha-\beta)}{2}$$

$$= \frac{1-\cos\alpha\cos\beta-\sin\alpha\sin\beta}{2}$$

$$= \frac{1-\dfrac{11}{61}\times\dfrac{3}{5}-\dfrac{60}{61}\times\dfrac{4}{5}}{2}$$

$$= \frac{305-33-240}{610}$$

$$= \frac{32}{610} = \frac{16}{305}$$

$$\cos^2\left(\frac{\alpha+\beta}{2}\right) = \frac{1+\cos(\alpha+\beta)}{2}$$

$$= \frac{1+\cos\alpha\cos\beta-\sin\alpha\sin\beta}{2}$$

$$= \frac{1+\dfrac{11}{61}\cdot\dfrac{3}{5}-\dfrac{60}{61}\cdot\dfrac{4}{5}}{2} = \frac{305+33-240}{610}$$

$$= \frac{98}{610} = \frac{49}{305}$$

Find the proper signs to be applied to the formula below:

43: $2\cos\dfrac{A}{2} = \pm\sqrt{1-\sin A} \pm \sqrt{1+\sin A}$, **when** $\dfrac{A}{2} = 278°$

$$\sin 278° = \sin(360-82°) = -\sin 82° = -\cos 8° < 0$$
$$\cos 278° = \cos(360-82°) = \cos 82° = \sin 8° > 0$$
$$8° < 45° \Rightarrow \cos 8° > \sin 8°$$

$$\cos\frac{A}{2} + \sin\frac{A}{2} = -\sqrt{1+\sin A}$$

$$\cos\frac{A}{2} - \sin\frac{A}{2} = +\sqrt{1-\sin A}$$

$$2\cos\frac{A}{2} = +\sqrt{1-\sin A} - \sqrt{1+\sin A}$$

If $A+B+C=\pi$ **, prove that:**

44: $\sin A + \sin B + \sin C = 4\cos\dfrac{A}{2}\cos\dfrac{B}{2}\cos\dfrac{C}{2}$

$$\text{LHS} = 2\sin\left(\frac{A+B}{2}\right)\cos\left(\frac{A-B}{2}\right) + 2\sin\frac{C}{2}\cos\frac{C}{2}$$

$$= 2\sin\left(\frac{\pi-C}{2}\right)\cos\left(\frac{A-B}{2}\right) + 2\cos\frac{C}{2}\sin\left(\frac{\pi-(A+B)}{2}\right)$$

$$= 2\sin\left(\frac{\pi}{2}-\frac{C}{2}\right)\cos\left(\frac{A-B}{2}\right) + 2\cos\frac{C}{2}\sin\left(\frac{\pi}{2}-\frac{A+B}{2}\right)$$

$$= 2\cos\frac{C}{2}\cos\left(\frac{A-B}{2}\right) + 2\cos\frac{C}{2}\cos\left(\frac{A+B}{2}\right)$$

$$= 2\cos\frac{C}{2}\left[\cos\left(\frac{A+B}{2}\right) + \cos\left(\frac{A-B}{2}\right)\right]$$

$$= 4\cos\frac{C}{2}\cos\frac{A}{2}\cos\frac{B}{2} = \text{RHS}$$

10 Solutions of triangles

In this chapter, we will take a look at a trigonometric ratios in terms of the sides of the triangles. This will also give us the insights in handling trigonometric ratios while handling triangles that are not right angled triangles.

1. Convention: In any $\triangle ABC$, the sides opposite to $\angle A$, $\angle B$ and $\angle C$ are represented by a, b and c.

2. In other words, a is the length of side BC. Similarly, b is the length of side AC and c is the length of side AB.

3. We use the symbol s to indicate the semi-perimeter of the triangle. In other words, semi-perimeter refers to half the perimeter of the triangle.

4. Clearly $s = \dfrac{a+b+c}{2}$.

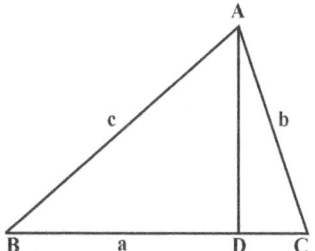

Fig: These two triangles can be used to get a grasp of all concepts in this chapter. Angle C is acute in the triangle above, while it is obtuse in the triangle below.

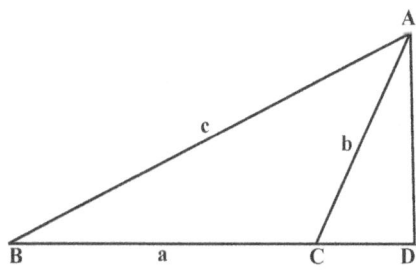

5. In any $\triangle ABC$, $\dfrac{\sin A}{a} = \dfrac{\sin B}{b} = \dfrac{\sin C}{c}$

   a. Consider the diagram below. AD is perpendicular to BC

   b. In $\triangle ABD$, $\dfrac{AD}{AB} = \sin BAD = c \sin B$

   c. In $\triangle ACD$, $\dfrac{AD}{AC} = \sin CAD = b \sin C$

   d. Equating these two values of AD, we have
       $c \sin B = b \sin C$

   e. $\dfrac{\sin B}{b} = \dfrac{\sin C}{c} \ldots (1)$

   f. Similarly by drawing a perpendicular B to AC, and going through similar steps as shown above, we can conclude that $a \sin C = c \sin A$

   g. $\dfrac{\sin C}{c} = \dfrac{\sin A}{a} \ldots (2)$

   h. Therefore, $\dfrac{\sin A}{a} = \dfrac{\sin B}{b} = \dfrac{\sin C}{c}$

   i. In other words, the sines of angles of a triangle are proportional to the opposite sides.

   j. Looking at the $\triangle$ s one more time, we can see that
       $$BD = c \cos B; CD = b \cos C$$

   k. Therefore $a = BC = BD + DC = c \cos B + b \cos C$

   l. If $\angle C$ is obtuse: $CD = b \cos(180^0 - C) = -b \cos C$

   m. $a = BC = BD - CD = c \cos B - (-\cos C)$
       $= c \cos B + b \cos C$

   n. We can summarize our findings as shown:

       i. $a = c \cos B + b \cos C$

       ii. $b = c \cos A + a \cos C$

       i. $c = b \cos A + a \cos B$

6. Now, let us turn our attention to cosines.

    a. $AB^2 = BC^2 + AC^2 - 2.BC.CD$

        And $CD = b\cos C$

        Therefore, $c^2 = a^2 + b^2 - 2ab\cos C$

        Or $\cos C = \dfrac{a^2 + b^2 - c^2}{2ab}$

        Similarly, the other equalities can be deduced.

    b.   Let us now summarize the findings on cosines.

        i. $\cos A = \dfrac{b^2 + c^2 - a^2}{2bc}$

        ii. $\cos B = \dfrac{a^2 + c^2 - b^2}{2ac}$

        iii. $\cos C = \dfrac{a^2 + b^2 - c^2}{2ab}$

    c.   We have considered the case where the angles are acute. If one of the angles was obstuse, we would have proceeded as below. We have assumed $\angle A$ is obtuse.

        i. $AB^2 = BC^2 + AC^2 + 2.BC.CD$

        ii. $CD = b\cos ACD = b\cos(180^0 - C) = -b\cos C$

        iii.   The equation now is the same as before. The other steps in the manipulation remain the same.

7. Now, let us turn our attention to the next set of derivations. Here, we will focus our attention on sines of half angles in terms of the sides.

    a. $\cos A = \dfrac{b^2 + c^2 - a^2}{2bc}$

    b.   We know, $\cos A = 1 - 2\sin^2\dfrac{A}{2}$

c. $2\sin^2\dfrac{A}{2} = 1 - \cos A$

$$= 1 - \frac{b^2 + c^2 - a^2}{2bc}$$

$$= \frac{2bc - b^2 - c^2 + a^2}{2bc}$$

$$= \frac{a^2 - (b-c)^2}{2bc}$$

$$= \frac{[a+(b-c)][a-(b-c)]}{2bc} \quad \ldots \qquad\qquad (1)$$

d.   We know, that semi-perimeter $s = \dfrac{a+b+c}{2}$,   or

$2s = a+b+c$

Therefore, $a+b-c = a+b+c-2c = 2(s-c)$

Similarly, $a-b+c = 2(s-b)$

e.   Substituting these values in equation (1), we have

$$2\sin^2\frac{A}{2} = \frac{2(s-c) \times 2s(s-b)}{2bc}$$

$$= \frac{2(s-b)(s-c)}{bc}$$

Therefore, $\sin\dfrac{A}{2} = \sqrt{\dfrac{(s-b)(s-c)}{bc}}$.

f.   We can now summarise our findings.

    i. $\sin\dfrac{A}{2} = \sqrt{\dfrac{(s-b)(s-c)}{bc}}$.

    ii. $\sin\dfrac{B}{2} = \sqrt{\dfrac{(s-a)(s-c)}{ac}}$.

iii. $\sin\dfrac{C}{2} = \sqrt{\dfrac{(s-a)(s-b)}{ab}}$.

g. By substituting $\cos A = 2\cos^2\dfrac{A}{2} - 1$, we have

$$\cos^2\frac{A}{2} = 1 + \cos A$$

$$= 1 + \frac{b^2 + c^2 - a^2}{2bc}$$

$$= \frac{2bc + b^2 + c^2 - a^2}{2bc}$$

$$= \frac{(b+c)^2 - a^2}{2bc}$$

$$= \frac{(b+c+a)(b+c-a)}{2bc}$$

$$= \frac{2s \times 2(s-a)}{2bc}$$

$$= \frac{s(s-a)}{bc}$$

Therefore, $\cos\dfrac{A}{2} = \sqrt{\dfrac{s(s-a)}{bc}}$

h. (f) We can now summarize our findings as follows.

i. i. $\cos\dfrac{A}{2} = \sqrt{\dfrac{s(s-a)}{bc}}$

j. ii. $\cos\dfrac{B}{2} = \sqrt{\dfrac{s(s-b)}{ac}}$

k. iii. $\cos\dfrac{A}{2} = \sqrt{\dfrac{s(s-c)}{ab}}$

8. In order to find tangents in terms of the half angles, we simply use

   the fact that $\tan\dfrac{A}{2} = \dfrac{\sin\dfrac{A}{2}}{\cos\dfrac{A}{2}}$

   a. $\tan\dfrac{A}{2} = \dfrac{\sqrt{\dfrac{(s-b)(s-c)}{bc}}}{\sqrt{\dfrac{s(s-a)}{bc}}}$

   $= \sqrt{\dfrac{(s-b)(s-c)}{s(s-a)}}$

   b. We can extend this to other half angles as well.

   i. $\tan\dfrac{A}{2} = \sqrt{\dfrac{(s-b)(s-c)}{s(s-a)}}$

   ii. $\tan\dfrac{B}{2} = \sqrt{\dfrac{(s-a)(s-c)}{s(s-b)}}$

   iii. $\tan\dfrac{C}{2} = \sqrt{\dfrac{(s-a)(s-b)}{s(s-c)}}$

   c. We know that $\sin A = 2\sin\dfrac{A}{2}\cos\dfrac{A}{2}$

   d. Subtituting the values for $\sin\dfrac{A}{2}$ and $\cos\dfrac{A}{2}$, we have the following results.

   i. i. $\sin A = \dfrac{2}{bc}\sqrt{s(s-a)(s-b)(s-c)}$

   ii. Similarly, $\sin b = \dfrac{2}{ac}\sqrt{s(s-a)(s-b)(s-c)}$

   iii. And, $\sin C = \dfrac{2}{ab}\sqrt{s(s-a)(s-b)(s-c)}$

10.1 Exercises

45:   **Given** $a = 125, b = 123,$ **and** $c = 62$ **. Find the sines of half the angles and the sines of the angles.**

$$s = \frac{a+b+c}{2} = 155; \; s-a = 30, s-b = 32, s-c = 93$$

$$\sin\frac{A}{2} = \sqrt{\frac{(s-b)(s-c)}{bc}} = \sqrt{\frac{32\times 93}{123\times 62}} = \frac{4}{\sqrt{41}}$$

$$\sin\frac{B}{2} = \sqrt{\frac{(s-a)(s-c)}{ac}} = \sqrt{\frac{30\times 93}{125\times 62}} = \frac{3}{5}$$

$$\sin\frac{C}{2} = \sqrt{\frac{(s-a)(s-b)}{ab}} = \sqrt{\frac{30\times 32}{125\times 123}} = \frac{8}{5\sqrt{41}}$$

$$\sin A = \frac{2}{bc}\sqrt{s(s-a)(s-b)(s-c)}$$

$$= \frac{2}{123\times 62}\sqrt{155\times 30\times 32\times 93} = \frac{40}{41}$$

$$\sin B = \frac{2}{ac}\sqrt{s(s-a)(s-b)(s-c)} = \frac{7440}{125\times 62} = \frac{120}{125} = \frac{24}{25}$$

$$\sin C = \frac{2}{ab}\sqrt{s(s-a)(s-b)(s-c)} = \frac{7440}{125\times 62} = \frac{496}{1025}$$

46:   **Given** $a = 287, b = 816$ **and** $c = 865,$ **find the values of** $\tan\dfrac{A}{2}$ **and** $\tan A$

$$s = \frac{a+b+c}{2} = 984; s-a = 697, s-b = 168, s-c = 119$$

$$\tan\frac{A}{2} = \sqrt{\frac{(s-b)(s-c)}{s(s-a)}} = \sqrt{\frac{168\times 119}{984\times 697}} = \frac{7}{41}$$

$$\tan A = \dfrac{2\tan\dfrac{A}{2}}{1-\tan^2\dfrac{A}{2}} = \dfrac{\dfrac{14}{41}}{1-\dfrac{49}{1681}} = \dfrac{287}{816}$$

In a triangle ABC, prove that:

47: $\quad b^2 \sin 2C + c^2 \sin 2B = 2bc \sin A$

We have:

$$\frac{\sin A}{a} = \frac{\sin B}{b} = \frac{\sin C}{c} = r$$

LHS $= b^2 \sin 2C + c^2 \sin 2B$

$= 2b^2 \sin c \cos C + 2c^2 \sin B \cos B$

$= 2b^2 (cr)\cos C + 2c^2 (br)\cos B$

$= 2bcr\left(\cos C + c\cos B\right) = 2bcra = 2bc \sin A \ = \text{RHS}$

48: $\quad a^2 + b^2 + c^2 = 2(bc\cos A + ca\cos B + ab\cos C)$

LHS

$$= \frac{2ab}{2ab}(a^2+b^2-c^2) + \frac{2bc}{2bc}(b^2+c^2-a^2) + \frac{2ac}{2ac}(c^2+a^2-b^2)$$

$= 2ab\cos C + 2bc\cos A + 2ac\cos B$

$= 2(bc\cos A + ac\cos B + ab\cos C) = \text{RHS}$

49:   **In a triangle whose sides are $3, 4$ and $\sqrt{38}$ cm respectively, prove that the largest angle is greater than $120°$.**

Biggest angle is opposite to the biggest side

$\qquad C$ is biggest angle

$$\cos C = \frac{a^2 + b^2 - c^2}{2ab} = \frac{9 + 16 - 38}{2 \times 3 \times 4} = \frac{-13}{24} = -\frac{1}{2} - \frac{1}{24}$$

$$\cos \frac{2\pi}{3} = -\frac{1}{2}$$

Hence:

$$C > \frac{2\pi}{3}$$

11 Heights and Distances

In this chapter, we will discuss a few scenarios for looking at problems in heights and distances. The basic strategy is one of representing the given information in form of a right angled triangle; assigning the values of known quantities and solving for the unknown using the trigonometric ratio and related identities/

1. **Scenario 1:**

   a. The angle of elevation is the angle formed by a horizontal line, and the line of sight looking up from the horizontal.
   b. The horizontal line under discussion may be real or imagined.

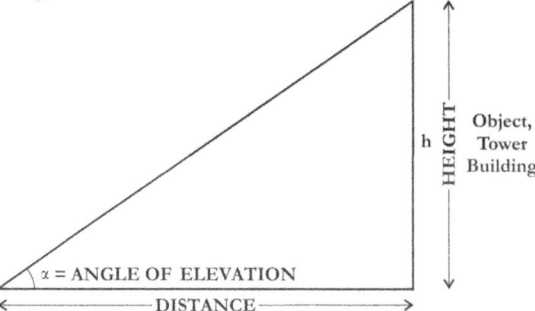

   c. We formulate the problem as a right angle.
   d. We can use the pythogorean theorem to solve for the unknown quantity.

2. **Scenario 2:**

   a. The angle of depression is the angle formed by the horizontal and the line of sight, looking down at the target from a certain elevation.
   b. As before, the horizontal may be real or imagined.

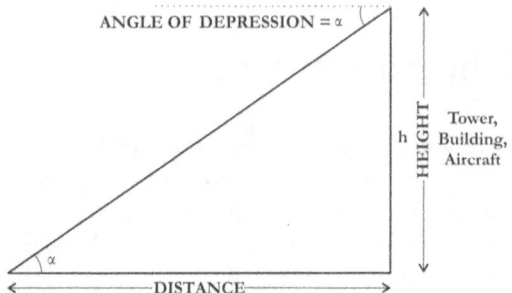

c. We formulate the representation as a right triangle.

d. We can use the Pythogorean theorem to solve for the unknown quantity.

3. Scenario 3:

a. When two angles of elevation are given at a specified distance apart, the problem can still be formulated as two right triangles sharing a common elevation and a portion of the horizontal.

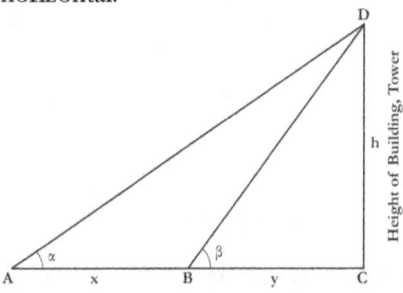

α,β Angle of Elevation from two points A and B

b. These elevation changes can occur in two ways.

i. The observer moves towards the reference elevation resulting in an increase in angle of elevation. This results in a decrease in the horizontal by a delta amount.

ii. When the observer moves away from the elevation, the angle of elevation reduces. This results in an increase in the horizontal by a delta amount.

c. We can use the pythogorean theorem to solve for the unknown quantity.

12 Inverse Circular Functions

1. We will now briefly touch upon the topic of inverse circular functions.

2. Let us assume $\sin\theta = a$. Then, $\sin^{-1}a = \theta$,means θ is the smallest angles, postive or negative for which a is the value of the ratio $\sin\theta$.

3. It must be carefully noted that $\sin^{-1}a$ is an angle and denotes the smallest numerical value of that angle.

4. We can similarly define $\cos^{-1}a$, $\tan^{-1}a$, $\sec^{-1}a$, $\csc^{-1}a$ and $\cot^{-1}a$.

5. The values of $\sin^{-1}a$ and $\tan^{-1}a$ always lie between 0^0 and 90^0, when positive and between -90^0 and 0^0 when negative. In other words, the values of $\sin^{-1}a$ and $\tan^{-1}a$, and therefore $\csc^{-1}a$ and $\cot^{-1}a$, always lies between -90^0 and 90^0.

6. The values of $\cos^{-1}a$, and therefore $\sec^{-1}a$ always lie between 0^0 and 180^0

7. The $\sin^{-1}a$ is the inverse sine function.

8. This is read as "sine minus one a" or "sine inverse a".

9. This is not to be confused with $(\sin\theta)^{-1} = 1/\sin\theta$. The -1 represents the inverse function and not the exponent.

10. In order to circumvent any confusion or ambiguity, several texts refer to $\sin^{-1}a$ as **arc sin a**.

12.1 Exercises

Prove the following:

50: $\cos^{-1}\dfrac{4}{5} + \tan^{-1}\dfrac{3}{5} = \tan^{-1}\dfrac{27}{11}$

Let:

$$\alpha = \cos^{-1}\frac{4}{5} \Rightarrow \tan\alpha = \frac{3}{4}$$

$$\beta = \tan^{-1}\frac{3}{5} \Rightarrow \tan\beta = \frac{3}{5}$$

$$\tan(\alpha+\beta) = \frac{\tan\alpha+\tan\beta}{1-\tan\alpha\tan\beta} = \frac{\dfrac{3}{4}+\dfrac{3}{5}}{1-\dfrac{3}{4}\times\dfrac{3}{5}} = \frac{27}{11}$$

$$\alpha+\beta = \tan^{-1}\frac{27}{11}$$

51: $\tan^{-1}\dfrac{1}{7} + \tan^{-1}\dfrac{1}{13} = \tan^{-1}\dfrac{2}{9}$

$$\text{LHS} = \tan^{-1}\frac{1}{7}\tan^{-1}\frac{1}{13} = \tan^{-1}\left(\frac{\dfrac{1}{7}+\dfrac{1}{13}}{1-\dfrac{1}{7}\cdot\dfrac{1}{13}}\right) = \tan^{-1}\frac{20}{90}$$

$$= \tan^{-1}\frac{2}{9} = \text{RHS}$$

52: $\tan^{-1}\dfrac{m}{n} - \tan^{-1}\dfrac{m-n}{m+n} = \dfrac{\pi}{4}$

$$\text{LHS} = \tan^{-1}\left(\dfrac{\dfrac{m}{n} - \dfrac{m-n}{m+n}}{1 + \dfrac{m}{n}\dfrac{(m-n)}{(m+n)}}\right) = \tan^{-1}\left(\dfrac{m(m+n) - n(m-n)}{n(m+n) + m(m-n)}\right)$$

$$= \tan^{-1}\left(\dfrac{m^2 + n^2}{m^2 + n^2}\right) = \tan^{-1}1 = \dfrac{\pi}{4} = \text{RHS}$$

53: $\tan^{-1}x = 2\tan^{-1}\left[\operatorname{cosec}\tan^{-1}x - \tan\cot^{-1}x\right]$

Let:

$$\theta = \tan^{-1}x \Rightarrow x = \tan\theta$$

$$\cot^{-1}x = \dfrac{\pi}{2} - \theta$$

RHS

$$= 2\tan^{-1}\left(\operatorname{cosec}\theta - \tan\left(\dfrac{\pi}{2} - \theta\right)\right) = 2\tan^{-1}\left(\dfrac{1}{\sin\theta} - \dfrac{\cos\theta}{\sin\theta}\right)$$

$$= 2\tan^{-1}\left(\dfrac{1 - \cos\theta}{\sin\theta}\right) = 2\tan^{-1}\left(\dfrac{2\sin^2\dfrac{\theta}{2}}{2\sin\dfrac{\theta}{2}\cos\dfrac{\theta}{2}}\right)$$

$$= 2\tan^{-1}\tan\dfrac{\theta}{2} = 2\times\dfrac{\theta}{2} = \theta = \tan^1 x = \text{LHS}$$

13 Problems

13.3 Measurement of Angles

Express in terms of a right angle the angles:

54: 60°

55: 63° 17' 25"

56: 130° 30'

57: 210° 30' 30"

58: 370° 20' 48"

Express in grades, minutes and seconds the angles:

59: 30°

60: 81°

61: 138° 30'

62: 235° 12' 36"

63: 475° 13' 48"

Express in terms of right angles, and also in degrees, minutes and seconds the angles:

64: 120^g

65: 39^g 45' 36"

66: 255^g 8' 9"

67: 759^g 0' 5"

Mark the position of the revolving line when it has traced out the following angles:

68: $\dfrac{4}{3}$ right angles

69:     $3\dfrac{1}{2}$ right angles

70:     $13\dfrac{1}{3}$ right angles

71:     120°

72:     315°

73:     745°

74:     1185°

75:     420^g

76:     875^g

77:     How many degrees, minutes and seconds are respectively passed over in $11\ \dfrac{1}{9}$ minutes by the hour and minute hands of a watch?

78:     The number of degrees in one acute angle of a right-angled triangle is equal to the number of grades in the other; express both the angles in degrees.

79:     Prove that the number of Sexagesimal minutes in any angle is to the number of Centesimal minutes in the same angle as 27 : 50.

80:     Divide 44°8' into two parts such that the number of Sexagesimal seconds in one part may be equal to the number of Centesimal seconds in the other part.

Solve the following:

81:     If the radius of the earth be 6400 km., what is the length of its circumference?

82:     The wheel of a railway carriage is 90 cm in diameter and makes 3 revolutions in a second; how fast is the train going?

83:     A mill sail whose length is 540 cm. makes 10 revolutions per minute. What distance does its end travel in an hour?

84:     The diameter of a halfpenny is an inch; what is the length of a piece of string which would just surround its curved edge?

85: Assuming that the earth describes in one year a circle, of 1, 49,700 Km. radius, whose center is the sun, how many miles; does the earth travel in a year?

86: The radius of a carriage wheel is 50 cm, and in 1/9th of a second it turns through 80° about its centre, which is fixed; how many km. does a point on the rim of the wheel travel in one hour?

Express in degrees, minutes and seconds the angles:

87: $\dfrac{\pi^c}{3}$

88: $\dfrac{4\pi^c}{3}$

89: $10\pi^c$

90: 1^c

91: 8^c

Express in grades, minutes and seconds the angles:

92: $\dfrac{4\pi^c}{5}$

93: $\dfrac{7\pi^c}{6}$

94: $10\pi^c$

Express in radians the following angles :

95: 60°

96: 110° 30'

97: 175° 45'

98: 47° 25' 36"

99: 395°

100: 60^g

101: $345^g\ 25'\ 36"$

102:    The difference between the two acute angles of a right-angled triangle is $\dfrac{2\pi}{5}$ radians; express the angles in degrees.

103:    One angle of a triangle is $\dfrac{2x}{3}$ grades and another is $\dfrac{3}{2}x$ degrees, whilst the third is $\dfrac{\pi x}{75}$ radians; express them all in degrees.

104:    The circular measure of two angles of a triangle are respectively 1/2 and 1/3; what is the number of degrees in the third angle?

105:    The angles of a triangle are in A.P. and the number of degrees in the least is to the number of radians in the greatest as 60 to π; find the angles in degrees.

106:    The angles of a triangle are in A.P. and the number of radians in the least angle is to the number of degrees in the mean angle as 1 : 120. Fine the angles in radians.

107:    Find the magnitude, in radians and degrees, of the interior angle of (1) A regular pentagon, (2) a regular heptagon, (3) a regular octagon, (4) a regular dodecagon, and (5) a regular polygon of 17 sides.

108:    The number of sides in two regular polygons are as 5 : 4, and the difference between their angles is 9°; find the number of sides in the polygons.

109:    Find two regular polygons such that the number of their sides may be as 3 to 4 and the number of degrees in an angle of the first to the number of grades in an angle of the second as 4 to 5.

110:    The angles of a quadrilateral are in A.P. and the greatest is double the least; express the least angle in radians.

111:    Find in radians, degrees,, and grades the angle between the hour-hand and the minute-hand of a clock at (1) half-past three, (2) twenty minutes to six, (3) a quarter past eleven.

112:    Find the times (1) between four and five o' clock when the angle between the minute-hand and the hour-hand is 78°, (2) between seven and eight o' clock when this angle is 54°.

113:    Find the number of degrees subtended at the 0 by arc those length in 0.357 times the radius.

114:    Express in radians and degrees the angle subtended at the center of a circle by an arc whose length is 15 cm, the radius being 25 cm.

115:    The value of the division on outer run of a graduated circle is 5'. The distance between successive graduations is 0.1 cm. Find the radius of the circle.

116:    The diameter of graduated circle is 72 cm. The graduations are 5' apart. Find the distance between the graduations.

117:    Find the radius of the globe which is such that the distance between two places on same meridian whose latitudes differs by $1°10'$ may be 5cm.

118:    Taking the radius of the earth as 6400 km. Find the difference in latitude of two places 100 km north of the other.

119:    Assuming the earth to be a sphere and the distance between two parallels of latitude which subtends an angle of $1°$ at the earth's center to be $69\frac{1}{9}$ km, what is the radius of the earth?

120:    The radius of a circle is 30 cm. Find the approx. length of arc if length of chord is of arc is $30°$.

121:    What is the ratio of radii of two circles at the center of which two arcs of the same length subtend angles of $60°$ .and $75°$?

122:    If an arc of length 10 cm, on a circle of 8 cm diameter subtends $143°14'22'$ at the center of the circle, find the value of π to 4 decimals.

123:    If the circumference of a circle be divided into 5 parts which are in AP; if the greatest part is 6 times the least; find in radians, the magnitude of the angles subtended at the center of circle.

124:    The perimeter of a certain sector of a circle is equal to the length of the arc of a semicircle having the same radius. Express the angle of the sector is degrees, minutes, sec.

125:    At what distance does a man whose height is 2 m, subtend an angle of 10'.

126:    Find the length which at a distance of 5280 m will subtend an angle of $1'$ at the eye.

127: Find the approximate distance at which a globe, $5\frac{1}{2}$ cm. in diameter, will subtend an angle of $6'$

128: Find the approximate distance of a tower whose height in 51m and which subtends and angle of $5\frac{5'}{11}$

129: A church spire, whose height in known to be 10 m subtends an angle of $9'$ at the eye. Find the approximate distance.

130: The radius of the earth being taken as 6400 km, and the distance of the moon from the earth being 60 times the radius of the earth, find the radius of moon if it subtends 16^1

131: When the moon is setting at any given place, the angle that is subtended at its center by the radius of the earth passing through the given place is 57'. If the radius of earth is 6400 kilometers, find the approx. distance to moon.

13.4 Basic Trigonometric Ratios

Prove the following statements:

1: $\cos^4 A - \sin^4 A + 1 = 2\cos^2 A$

2: $(\sin A + \cos A)(1 - \sin A \cos A) = \sin^3 A + \cos^3 A$

3: $\dfrac{\sin A}{1 + \cos A} + \dfrac{1 + \cos A}{\sin A} = 2\operatorname{cosec} A$

4: $\cos^6 A + \sin^6 A = 1 - 3\sin^2 A \cos^2 A$

5: $\sqrt{\dfrac{1 - \sin A}{1 + \sin A}} = \sec A - \tan A$

6: $\dfrac{\operatorname{cosec} A}{\operatorname{cosec} A - 1} + \dfrac{\operatorname{cosec} A}{\operatorname{cosec} A + 1} = 2\sec^2 A$

7: $(\sec A + \cos A)(\sec A - \cos A) = \tan^2 A + \sin^2 A$

8:
$$\frac{1}{\cot A + \tan A} = \sin A \cos A$$

9:
$$\frac{1}{\sec A - \tan A} = \sec A + \tan A$$

10:
$$\frac{1 - \tan A}{1 + \tan A} = \frac{\cot A - 1}{\cot A + 1}$$

11:
$$\frac{1 + \tan^2 A}{1 + \cot^2 A} = \frac{\sin^2 A}{\cos^2 A}$$

12:
$$\frac{\sec A - \tan A}{\sec A + \tan A} = 1 - 2 \sec A \tan A + 2 \tan^2 A$$

13:
$$\frac{\tan A}{1 - \cot A} + \frac{\cot A}{1 - \tan A} = \sin A \operatorname{cosec} A + 1$$

14:
$$\frac{\cos A}{1 - \tan A} + \frac{\sin A}{1 - \cot A} = \sin A + \cos A$$

15:
$$(\sin A + \cos A)(\cot A + \tan A) = \sec A + \operatorname{cosec} A$$

16:
$$\cot^4 A + \cot^2 A = \operatorname{cosec}^4 A - \operatorname{cosec}^2 A$$

17:
$$\sqrt{\operatorname{cosec}^2 A - 1} = \cos A \operatorname{cosec} A$$

18:
$$\sec^2 A \operatorname{cosec}^2 A = \tan^2 A + \cot^2 A + 2$$

19:
$$\tan^2 A - \sin^2 A = \sin^4 A \sec^2 A$$

20:
$$(1 + \cot A - \operatorname{cosec} A)(1 + \tan A + \sec A) = 2$$

21:
$$\frac{\cot A \cos A}{\cot A + \cos A} = \frac{\cot A - \cos A}{\cot A \cos A}$$

22:
$$\frac{\cot A + \tan B}{\cot B + \tan A} = \cot A \tan B$$

23:
$$\left(\frac{1}{\sec^2 \alpha - \cos^2 \alpha} + \frac{1}{\operatorname{cosec}^2 \alpha - \sin^2 \alpha} \right) \cos^2 \alpha \sin^2 \alpha$$
$$= \frac{1 - \cos^2 \alpha \sin^2 \alpha}{2 + \cos^2 \alpha \sin^2 \alpha}$$

24:     $\sin^8 A - \cos^8 A = (\sin^2 A - \cos^2 A)(1 - 2\sin^2 A \cos^2 A)$

25:     $\dfrac{\cos A \operatorname{cosec} A - \sin A \sec A}{\cos A + \sin A} = \operatorname{cosec} A - \sec A$

26:     $\dfrac{\tan A + \sec A - 1}{\tan A - \sec A + 1} = \dfrac{1 + \sin A}{\cos A}$

27:     $(\tan \alpha + \operatorname{cosec} \beta)^2 - (\cot \beta - \sec \alpha)^2$
$$= 2 \tan \alpha \cot \beta (\operatorname{cosec} \alpha + \sec \beta)$$

28:     $2\sec^2 \alpha - \sec^4 \alpha - 2\operatorname{cosec}^2 \alpha + \operatorname{cosec}^4 \alpha = \cot^4 \alpha - \tan^4 \alpha$

29:     $(\sin \alpha + \operatorname{cosec} \alpha)^2 + (\cos \alpha + \sec \alpha)^2 = \tan^2 \alpha + \cot^2 \alpha + 7$

30:     $(\operatorname{cosec} A + \cot A)\operatorname{covers} A - (\sec A + \tan A)\operatorname{vers} A$
$$= (\operatorname{cosec} A - \sec A)(2 - \operatorname{vers} A \operatorname{covers} A)$$

31:     $(1 + \cot A + \tan A)(\sin A - \cos A) = \dfrac{\sec A}{\operatorname{cosec}^2 A} - \dfrac{\operatorname{cosec} A}{\sec^2 A}$

32:     $2\operatorname{vers} A + \cos^2 A = 1 + \operatorname{vers}^2 A$

33:     Express all the other trigonometrical ratios in terms of the cosine.

34:     Express all the ratios in terms of the cosecant.

35:     Express all the ratios in terms of the secant.

36:     The sine of a certain angle is 1/4; find the numerical values of the other trigonometrical ratios of this angle.

37:     If $\sin \theta = \dfrac{12}{13}$, find $\tan \theta$ and vers θ

38:     If $\sin A = \dfrac{11}{61}$, find $\tan A$, $\cos A$ and $\sec A$.

39:     If $\cos \theta = \dfrac{4}{5}$, find $\sin \theta$ and $\cot \theta$.

40:     If $\cos A = \dfrac{9}{41}$, find $\tan A$ and $\operatorname{cosec} A$.

41: If $\tan\theta = \dfrac{3}{4}$, find the sine, cosine, versine, and cosecant of θ.

42: If $\cot\theta = \dfrac{15}{8}$, find $\cos\theta$ and $\operatorname{cosec}\theta$

43: If $\sec A = \dfrac{3}{2}$, find $\tan A$ and $\operatorname{cosec} A$

44: If $2\sin\theta = 2 - \cos\theta$, find $\sin\theta$

45: If $8\sin\theta = 4 + \cos\theta$, find $\sin\theta$

46: If $\tan\theta + \sec\theta = 1.5$, find $\sin\theta$

47: If $\cot\theta + \operatorname{cosec}\theta = 5$, find $\cos\theta$

48: If $3\sec^4\theta + 8 = 10\sec^2\theta$, find the values of $\tan\theta$

49: If $\tan^2\theta + \sec\theta = 5$, find $\cos\theta$.

50: If $\tan\theta + \cot\theta = 2$, find $\sin\theta$.

51: If $\tan\theta = \dfrac{2x(x+1)}{2x+1}$ find $\sin\theta$ and $\cos\theta$.

52: If A = 30°, verify that

   a. $\cos 2A = \cos^2 A - \sin^2 A = 2\cos^2 A - 1$

   b. $\sin 2A = 2\sin A \cos A$

   c. $\cos 3A = 4\cos^3 A - 3\cos A$

   d. $\sin 3A = 3\sin A - 4\sin^3 A$ and

   e. $\tan 2A = \dfrac{2\tan A}{1 - \tan^2 A}$

53: If A = 45°, verify that

   a. $\sin 2A = 2\sin A \cos A$

   b. $\cos 2A = 1 - 2\sin^2 A$

c. $\qquad \tan 2A = \dfrac{2\tan A}{1 - \tan^2 A}$

Verify that

54: $\quad \sin^2 30° + \sin^2 45° + \sin^2 60° = \dfrac{3}{2}$

55: $\quad \tan^2 30° + \tan^2 45° + \tan^2 60° = 4\dfrac{1}{3}$

56: $\quad \sin 30° \cos 60° + \cos 30° \sin 60° = 1$

57: $\quad \cos 45° \cos 60° - \sin 45° \sin 60° = \dfrac{\sqrt{3} - 1}{2\sqrt{2}}$

58: $\quad \dfrac{4}{3}\cot^2 30° + 3\sin^2 60° + 2\operatorname{cosec}^2 60° - \dfrac{3}{4}\tan^2 30° = 3\dfrac{1}{3}$

59: $\quad \operatorname{cosec}^2 45° \sec^2 30° \sin^2 90° \cos 60° = 1\dfrac{1}{3}$

60: $\quad 4\cot^2 45° - \sec^2 60° + \sin^3 30° = \dfrac{1}{8}$

13.6 Trigonometric functions of angles of any size and sign

Prove that:

1: $\sin 420° \cos 390° + \cos(-300°)\sin(-330°) = 1$

2: $\tan 225° \cot 405° + \tan 765° \cos 675° = 0$

What are the values of $\cos A - \sin A$ and $\tan A + \cot A$ when A has the values

3: $\dfrac{\pi}{3}$

4: $\dfrac{5\pi}{4}$

5: $\dfrac{7\pi}{4}$

6: $\dfrac{11\pi}{3}$

What values between $0°$ and $360°$ may A have when

7: $\sin A = \dfrac{1}{\sqrt{2}}$

8: $\tan A = -1$

9: $\cot A = -\sqrt{3}$

10: $\sec A = -\dfrac{2}{\sqrt{3}}$

11: $\operatorname{cosec} A = -2$

Express in terms of the ratios of a positive angle, which is less than 45°, the quantities

12:    $\sin(-65°)$

13:    $\cos(-84°)$

14:    $\sin 168°$

15:    $\cos 287°$

16:    $\tan(-246°)$

17:    $\sin 843°$

18:    $\cos(-928°)$

19:    $\tan 1145°$

20:    $\cos 1410°$

21:    $\cot(-1054°)$

22:    $\sec 1327°$

23:    $\operatorname{cosec}(-756°)$

What sign has $\sin A + \cos A$ for the following values of A?

24:    $140°$

25:    $-356°$

26:    $-1125°$

What sign has $\sin A - \cos A$ for the following values of A?

27:    $215°$

28:    $825°$

29:    $-634°$

30:    $-457°$

31:    Find the sines and cosines of all angles in the first four quadrants whose tangents are equal to $\cos 135°$

Prove that:

32: $\sin(270° + A) = -\cos A$

33: $\tan(270° + A) = -\cot A$

34: $\cos(270° - A) = -\sin A$

35: $\cot(270° - A) = \tan A$

36: $\cos A + \sin(270° + A) - \sin(270° - A) + \cos(180° + A) = 0$

37: $\sec(270° - A)\sec(90° - A)$
$\quad - \tan(270° - A)\tan(90° + A) + 1 = 0$

38: $\cot A + \tan(180° + A) + \tan(90° + A) + \tan(360° - A) = 0$

13.7 General expressions for all trigonometric ratios

What are the most general values of θ which satisfy the equations?

1:      $\sin\theta = \dfrac{1}{2}$

2:      $\sin\theta = -\dfrac{\sqrt{3}}{2}$

3:      $\sin\theta = \dfrac{1}{\sqrt{2}}$

4:      $\cos\theta = -\dfrac{1}{2}$

5:      $\cos\theta = \dfrac{\sqrt{3}}{2}$

6:      $\cos\theta = -\dfrac{1}{\sqrt{2}}$

7:      $\tan\theta = \sqrt{3}$

8:      $\tan\theta = -1$

9:      $\cot\theta = 1$

10:      $\sec\theta = 2$

11:      $\operatorname{cosec}\theta = \dfrac{2}{\sqrt{3}}$

12:      $\sin^2\theta = 1$

13:      $\cos^2\theta = \dfrac{1}{4}$

14:      $\tan^2\theta = \dfrac{1}{3}$

15:      $2\cot^2\theta = \operatorname{cosec}^2\theta$

16:    $\sec^2\theta = \dfrac{4}{3}$

17:    What is the most general value of θ that satisfies both of the equations $\cos\theta = -\dfrac{1}{\sqrt{2}}$ and $\tan\theta = 1$?

18:    What is the most general value of θ that satisfies both of the equations $\cot\theta = -\sqrt{3}$ and $\operatorname{cosec}\theta = -2$?

19:    If $\cos(A-B) = \dfrac{1}{2}$ and $\sin(A+B) = \dfrac{1}{2}$, find the smallest positive values of A and B and also their most general values.

20:    If $\tan(A-B) = 1$, and $\sec(A+B) = \dfrac{2}{\sqrt{3}}$, find the smallest positive values of A and B and also their most general values.

21:    Find the angles between $0°$ and $360°$ which have respectively (a). their sines equal to $\dfrac{\sqrt{3}}{2}$, (b). their cosines equal to $-\dfrac{1}{2}$, and (c). their tangents equal to $\dfrac{1}{\sqrt{3}}$

22:    Taking into consideration only angles between $0°$ and $180°$, how many values of x are there if (a). $\sin x = \dfrac{5}{7}$ (b). $\cos x = \dfrac{1}{5}$, (c). $\cos x = -\dfrac{4}{5}$, (d). $\tan x = \dfrac{2}{3}$ and (e). $\cot x = -7$?

23:    Show that the same angles are indicated by the two following formulae: (a). $(2n-1)\dfrac{\pi}{2} + (-1)^n \dfrac{\pi}{3}$, and (b). $2n\pi \pm \dfrac{\pi}{6}$, n being any integer.

24: Prove that the two formulae (a). $\left(2n+\dfrac{1}{2}\right)\pi\pm\alpha$ and (b).

$n\pi+(-1)^n\left(\dfrac{\pi}{2}-\alpha\right)$ denote the same angles, n being any integer.

25: If $\theta-\alpha=n\pi+(-1)^n\beta$, prove that $\theta=2m\pi+\alpha+\beta$ or else that $\theta=(2m+1)\pi+\alpha-\beta$, where m and n are any integers.

26: If $\cos p\theta+\cos q\theta=0$, prove that the different values of θ from two arithmetical progressions in which the common differences are $\dfrac{2\pi}{p+q}$ and $\dfrac{2\pi}{p-q}$ respectively.

Find the general value of θ for the following:

27: $\cos^2\theta-\sin\theta-\dfrac{1}{4}=0$

28: $2\sin^2\theta+3\cos\theta=0$

29: $\cos\theta+\cos^2\theta=1$

30: $4\cos\theta-3\sec\theta=2\tan\theta$

31: $\sin^2\theta-2\cos\theta+\dfrac{1}{4}=0$

32: $\tan^2\theta-(1+\sqrt{3})\tan\theta+\sqrt{3}=0$

33: $\cot^2\theta+\left(\sqrt{3}+\dfrac{1}{\sqrt{3}}\right)\cot\theta+1=0$

34: $\cot\theta-ab\tan\theta=a-b$

35: $\tan^2\theta+\cot^2\theta=2$

36: $\sec\theta-1=(\sqrt{2}-1)\tan\theta$

37: $3(\sec^2\theta+\tan^2\theta)=5$

38: $\cot\theta+\tan\theta=2\csc\theta$

39: $4\cos^2\theta+\sqrt{3}=2(\sqrt{3}+1)\cos\theta$

40: $\quad 3\sin^2\theta - 2\sin\theta = 1$

41: $\quad \sin 5\theta = \dfrac{1}{\sqrt{2}}$

42: $\quad \sin 9\theta = \sin\theta$

43: $\quad \sin 3\theta = \sin 2\theta$

44: $\quad \cos m\theta = \cos n\theta$

45: $\quad \cos 3\theta = \sin 2\theta$

46: $\quad \cos m\theta = \sin n\theta$

47: $\quad \cot\theta = \tan 8\theta$

48: $\quad \cot\theta = \tan n\theta$

49: $\quad \tan 2\theta = \tan\dfrac{2}{\theta}$

50: $\quad \tan 2\theta \tan\theta = 1$

51: $\quad \tan^2 3\theta = \cot^2\alpha$

52: $\quad \tan 3\theta = \cot\theta$

53: $\quad \tan^2 3\theta = \tan^2\alpha$

54: $\quad 3\tan^2\theta = 1$

55: $\quad \tan(\pi\cot\theta) = \cot(\pi\tan\theta)$

56: $\quad \sin(\theta-\phi) = \dfrac{1}{2}$ and $\cos(\theta+\phi) = \dfrac{1}{2}$

57: $\quad \cos(2x+3y) = \dfrac{1}{2}$ and $\cos(3x+2y) = \dfrac{\sqrt{3}}{2}$

58: $\quad$ Find all the angles between $0°$ and $90°$ which satisfy the equation $\sec^2\theta\,\mathrm{cosec}^2\theta + 2\,\mathrm{cosec}^2\theta = 8$

59: $\quad$ If the coversine of an angle be $\dfrac{1}{3}$, find its cosine and cotangent.

13.8 Trigonometric Ratio: Sum and difference of two angles

1: If $\sin\alpha = \dfrac{3}{2}$ and $\cos\beta = \dfrac{9}{41}$, find the value of $\sin(\alpha-\beta)$ and $\cos(\alpha+\beta)$

2: If $\sin\alpha = \dfrac{15}{17}$ and $\cos\beta = \dfrac{12}{13}$, find the values of $\sin(\alpha+\beta)$, $\cos(\alpha-\beta)$ and $\tan(\alpha+\beta)$

Prove the following:

3: $\cos(45° - A)\cos(45° - B) - \sin(45° - A)\sin(45° - B)$
$= \sin(A+B)$

4: $\sin(45° + A)\cos(45° - B) + \cos(45° + A)\sin(45° - B)$
$= \cos(A-B)$

5: $\dfrac{\sin(A-B)}{\cos A \cos B} + \dfrac{\sin(B-C)}{\cos B \cos C} + \dfrac{\sin(C-A)}{\cos C \cos A} = 0$

6: $\sin 105° + \cos 105° = \cos 45°$

7: $\sin 75° - \sin 15° = \cos 105° + \cos 15°$

8: $\cos\alpha\cos(\gamma-\alpha) - \sin\alpha\sin(\gamma-\alpha) = \cos\gamma$

9: $\cos(\alpha+\beta)\cos\gamma - \cos(\beta+\gamma)\cos\alpha = \sin\beta\sin(\gamma-\alpha)$

10: $\sin(n+1)A\sin(n+2)A + \cos(n+1)A\cos(n+2)A = \cos A$

Prove the following:

11: $\dfrac{\sin 7\theta - \sin 5\theta}{\cos 7\theta + \cos 5\theta} = \tan\theta$

12: $\dfrac{\cos 6\theta - \cos 4\theta}{\sin 6\theta + \sin 4\theta} = -\tan\theta$

13: $\dfrac{\sin A + \sin 3A}{\cos A + \cos 3A} = \tan 2A$

14: $\dfrac{\sin 7A - \sin A}{\sin 8A - \sin 2A} = \cos 4A \sec 5A$

15: $\dfrac{\sin 2A + \sin 2B}{\sin 2A - \sin 2B} = \dfrac{\tan(A+B)}{\tan(A-B)}$

16: $\dfrac{\sin A + \sin 2A}{\cos A - \cos 2A} = \cot \dfrac{A}{2}$

17: $\dfrac{\sin 5A - \sin 3A}{\cos 3A + \cos 5A} = \tan A$

18: $\dfrac{\cos 2B - \cos 2A}{\sin 2B + \sin 2A} = \tan(A-B)$

19: $\cos(A+B) + \sin(A-B) = 2\sin(45° + A)\cos(45° + B)$

20: $\dfrac{\cos 3A - \cos A}{\sin 3A - \sin A} + \dfrac{\cos 2A - \cos 4A}{\sin 4A - \sin 2A} = \dfrac{\sin A}{\cos 2A \cos 3A}$

21: $\dfrac{\sin(4A - 2B) + \sin(4B - 2A)}{\cos(4A - 2B) + \cos(4B - 2A)} = \tan(A+B)$

22: $\dfrac{\tan 5\theta + \tan 3\theta}{\tan 5\theta - \tan 3\theta} = 4\cos 2\theta \cos 4\theta$

23: $\dfrac{\cos 3\theta + 2\cos 5\theta + \cos 7\theta}{\cos \theta + 2\cos 3\theta + \cos 5\theta} = \cos 2\theta - \sin 2\theta \tan 3\theta$

24: $\dfrac{\sin A + \sin 3A + \sin 5A + \sin 7A}{\cos A + \cos 3A + \cos 5A + \cos 7A} = \tan 4A$

25: $\dfrac{\sin(\theta + \phi) - 2\sin \theta + \sin(\theta - \phi)}{\cos(\theta + \phi) - 2\cos \theta + \cos(\theta - \phi)} = \tan \theta$

26: $\dfrac{\sin A + 2\sin 3A + \sin 5A}{\sin 3A + 2\sin 5A + \sin 7A} = \dfrac{\sin 3A}{\sin 5A}$

27: $\dfrac{\sin(A - C) + 2\sin A + \sin(A + C)}{\sin(B - C) + 2\sin B + \sin(B + C)} = \dfrac{\sin A}{\sin B}$

28: $\dfrac{\sin A - \sin 5A + \sin 9A - \sin 13A}{\cos A - \cos 5A - \cos 9A + \cos 13A} = \cot 4A$

29: $\dfrac{\sin A + \sin B}{\sin A - \sin B} = \tan\dfrac{A+B}{2}\cot\dfrac{A-B}{2}$

30: $\dfrac{\sin A + \sin B}{\cos A + \cos B} = \tan\dfrac{A+B}{2}$

31: $\dfrac{\sin A - \sin B}{\cos B - \cos A} = \cot\dfrac{A+B}{2}$

32: $\dfrac{\begin{pmatrix}\cos(A+B+C)+\cos(-A+B+C)\\ +\cos(A-B+C)+\cos(A+B-C)\end{pmatrix}}{\begin{pmatrix}\sin(A+B+C)+\sin(-A+B+C)\\ -\sin(A-B+C)+\sin(A+B-C)\end{pmatrix}} = \cot B$

33: $\cos 3A + \cos 5A + \cos 7A + \cos 15A$
$$= 4\cos 4A\cos 5A\cos 6A$$

34: $\cos(-A+B+C)+\cos(A-B+C)+\cos(A+B-C)$
$$+\cos(A+B+C) = 4\cos A\cos B\cos C$$

35: $\sin 50° - \sin 70° + \sin 10° = 0$

36: $\sin\alpha + \sin 2\alpha + \sin 4\alpha + \sin 5\alpha = 4\cos\dfrac{\alpha}{2}\cos\dfrac{3\alpha}{2}\sin 3\alpha$

Simplify

37: $\cos\left\{\theta + \left(n - \dfrac{3}{2}\right)\phi\right\} - \cos\left\{\theta + \left(n + \dfrac{3}{2}\right)\phi\right\}$

38: $\sin\left\{\theta + \left(n + \dfrac{1}{2}\right)\phi\right\} + \sin\left\{\theta + \left(n + \dfrac{1}{2}\right)\phi\right\}$

Express as a sum or difference the following:

39:      $2\sin 5\theta \sin 7\theta$

40:      $2\cos 7\theta \sin 5\theta$

41:      $2\cos 11\theta \cos 3\theta$

42:      $2\sin 54° \sin 66°$

Prove that

43:      $\sin\dfrac{\theta}{2}\sin\dfrac{7\theta}{2} + \sin\dfrac{3\theta}{2}\sin\dfrac{11\theta}{2} = \sin 2\theta \sin 5\theta$

44:      $\cos 2\theta \cos\dfrac{\theta}{2} - \cos 3\theta \cos\dfrac{9\theta}{2} = \sin 5\theta \sin\dfrac{5\theta}{2}$

45:      $\sin A \sin(A+2B) - \sin B \sin(B+2A)$
$$= \sin(A-B)\sin(A+B)$$

46:      $(\sin 3A + \sin A)\sin A + (\cos 3A - \cos A)\cos A = 0$

47:      $\dfrac{2\sin(A-C)\cos C - \sin(A-2C)}{2\sin(B-C)\cos C - \sin(B-2C)} = \dfrac{\sin A}{\sin B}$

48:      $\dfrac{\sin A \sin 2A + \sin 3A \sin 6A + \sin 4A \sin 13A}{\sin A \cos 2A + \sin 3A \cos 6A + \sin 4A \cos 13A} = \tan 9A$

49:      $\dfrac{\cos 2A \cos 3A - \cos 2A \cos 7A + \cos A \cos 10A}{\sin 4A \sin 3A - \sin 2A \sin 5A + \sin 4A \sin 7A}$
$$= \cot 6A \cot 5A$$

50:      $\cos(36° - A)\cos(36° + A) + \cos(54° + A)\cos(54° - A)$
$$= \cos 2A$$

51:      $\cos A \sin(B-C) + \cos B \sin(C-A) + \cos C \sin(A-B) = 0$

52:      $\sin(45° + A)\sin(45° - A) = \dfrac{1}{2}\cos 2A$

53:      $\operatorname{vers}(A+B)\operatorname{vers}(A-B) = (\cos A - \cos B)^2$

54:      $\sin(\beta - \gamma)\cos(\alpha - \delta) + \sin(r - \alpha)\cos(\beta - \delta)$
$$+ \sin(\alpha - \beta)\cos(\gamma - \delta) = 0$$

55:     $2\cos\dfrac{\pi}{13}\cos\dfrac{9\pi}{13}+\cos\dfrac{3\pi}{13}+\cos\dfrac{5\pi}{13}=0$

Solve the following:

56:     If $\tan A=\dfrac{1}{2}$ and $\tan B=\dfrac{1}{3}$, find the values of $\tan(2A+B)$ and $\tan(2A-B)$

57:     If $\tan A=\dfrac{\sqrt{3}}{4-\sqrt{3}}$ and $\tan B=\dfrac{\sqrt{3}}{4+\sqrt{3}}$, prove that $\tan(A-B)=.375$

58:     If $\tan A=\dfrac{n}{n+1}$ and $\tan B=\dfrac{n}{2n+1}$, find $\tan(A+B)$.

59:     If $\tan\alpha=\dfrac{5}{6}$ and $\tan\beta=\dfrac{1}{11}$, prove that $\alpha+\beta=\dfrac{\pi}{4}$.

Prove that:

$\tan\left(\dfrac{\pi}{4}+\theta\right)\tan\left(\dfrac{3\pi}{4}+\theta\right)=-1$

60:     $\cot\left(\dfrac{\pi}{4}+\theta\right)\cot\left(\dfrac{\pi}{4}-\theta\right)=1$

61:     $1+\tan A\tan\dfrac{A}{2}=\tan A\cot\dfrac{A}{2}-1=\sec A$

13.9 Trigonometric Ratios: Multiple and sub-multiple angles

Solve the following:

1:    Find the value of $\sin 2\alpha$ when: (a). $\cos \alpha = \dfrac{3}{5}$, (b). $\sin \alpha = \dfrac{12}{13}$, and (c). $\tan \alpha = \dfrac{16}{63}$

2:    Find the value of $\cos 2\alpha$ when (a). $\cos \alpha = \dfrac{15}{17}$, (b). $\sin \alpha = \dfrac{4}{5}$, and (c.) $\tan \alpha = \dfrac{5}{12}$

3:    If $\tan \theta = \dfrac{b}{a}$, find the value of $a \cos 2\theta + b \sin 2\theta$

Prove that

4:    $\dfrac{\sin 2A}{1 + \cos 2A} = \tan A$

5:    $\dfrac{1 - \cos 2A}{1 + \cos 2A} = \tan^2 A$

6:    $\tan A + \cot A = 2 \operatorname{cosec} 2A$

7:    $\tan A - \cot A = -2 \cot 2A$

8:    $\operatorname{cosec} 2A + \cot 2A = \cot A$

9:    $\dfrac{1 - \cos A + \cos B - \cos(A + B)}{1 + \cos A - \cos B - \cos(A + B)} = \tan \dfrac{A}{2} \cot \dfrac{B}{2}$

10:    $\dfrac{\cos A}{1 \pm \sin A} = \tan \left(\dfrac{\pi}{4} \pm \dfrac{A}{2} \right)$

11: $$\frac{1+\tan^2(\frac{\pi}{4}-A)}{1-\tan^2(\frac{\pi}{4}-A)} = \operatorname{cosec} 2A$$

12: $$\frac{\sin\alpha+\sin\beta}{\sin\alpha-\sin\beta} = \frac{\tan\frac{\alpha+\beta}{2}}{\tan\frac{\alpha-\beta}{2}}$$

13: $$\frac{\sin^2 A-\sin^2 B}{\sin A\cos A-\sin B\cos B} = \tan(A+B)$$

14: $$\tan\left(\frac{\pi}{4}+\theta\right)-\tan\left(\frac{\pi}{4}-\theta\right) = 2\tan 2\theta$$

15: $$\frac{\cos A+\sin A}{\cos A-\sin A}-\frac{\cos A-\sin A}{\cos A+\sin A} = 2\tan 2A$$

16: $$\cot(A+15°)-\tan(A-15°) = \frac{4\cos 2A}{1+2\sin 2A}$$

17: $$\frac{\sin\theta+\sin 2\theta}{1+\cos\theta+\cos 2\theta} = \tan\theta$$

18: $$\frac{1+\sin\theta-\cos\theta}{1+\sin\theta+\cos\theta} = \tan\frac{\theta}{2}$$

19: $$\frac{\sin(n+1)A-\sin(n-1)A}{\cos(n+1)A+2\cos nA+\cos(n-1)A} = \tan\frac{A}{2}$$

20: $$\frac{\sin(n+1)A+2\sin nA+\sin(n-1)A}{\cos(n-1)A-\cos(n+1)A} = \cot\frac{A}{2}$$

21: $$\sin(2n+1)A\sin A = \sin^2(n+1)A-\sin^2 nA$$

22: $$\frac{\sin(A+3B)+\sin(3A+B)}{\sin 2A+\sin 2B} = 2\cos(A+B)$$

23: $$\sin 3A+\sin 2A-\sin A = 4\sin A\cos\frac{A}{2}\cos\frac{3A}{2}$$

24: $\quad \tan 2A = (\sec 2A + 1)\sqrt{\sec^2 A - 1}$

25: $\quad \cos^3 2\theta + 3\cos 2\theta = 4(\cos^6 \theta - \sin^6 \theta)$

26: $\quad 1 + \cos^2 2\theta = 2(\cos^4 \theta + \sin^4 \theta)$

27: $\quad \sec^2 A(1 + \sec 2A) = 2\sec 2A$

28: $\quad \operatorname{cosec} A - 2\cot 2A \cos A = 2\sin A$

Prove that

29: $\quad \cot A = \dfrac{1}{2}\left(\cot\dfrac{A}{2} - \tan\dfrac{A}{2}\right)$

30: $\quad \sin\alpha \sin(60° - \alpha)\sin(60° + \alpha) = \dfrac{1}{4}\sin 3\alpha$

31: $\quad \cos\alpha \sin(60° - \alpha)\cos(60° + \alpha) = \dfrac{1}{4}\cos 3\alpha$

32: $\quad \cot\alpha + \cos(60° + \alpha) - \cot(60° - \alpha) = 3\cot 3\alpha$

33: $\quad \cos 20° \cos 40° \cos 60° \cos 80° = \dfrac{1}{16}$

34: $\quad \sin 20° \sin 40° \sin 60° \sin 80° = \dfrac{3}{16}$

35: $\quad \cos 4\alpha = 1 - 8\cos^2 \alpha + 8\cos^4 \alpha$

36: $\quad \sin 4\alpha = 4\sin\alpha \cos^3 \alpha - 4\cos\alpha \sin^3 \alpha$

37: $\quad \cos 6\alpha = 32\cos^6 \alpha - 48\cos^4 \alpha + 18\cos^2 \alpha - 1$

38: $\quad \tan 3A \tan 2A \tan A = \tan 3A - \tan 2A - \tan A$

39: $\quad \dfrac{2\cos 2^n\theta + 1}{2\cos\theta + 1} = (2\cos\theta - 1)(2\cos 2\theta - 1)\cdots(2\cos 2^{n-1}\theta - 1)$

Solve the following:

40: $\quad$ If $\sin\theta = \dfrac{1}{2}$ and $\sin\phi = \dfrac{1}{3}$, find the values of $\sin(\theta + \phi)$ and $\sin(2\theta + 2\phi)$

41:     The tangent of an angle is 2.4. Find its cosecant, the cosecant of half the angle, and the cosecant of the supplement of double the angle.

42:     If $\cos \alpha = \dfrac{3}{5}$ and $\cos \beta = \dfrac{4}{5}$, find the value of $\cos \dfrac{\alpha - \beta}{2}$, the angles α and β being positive acute angles.

43:     Given $\sec \theta = 1\dfrac{1}{4}$, find $\tan \dfrac{\theta}{2}$ and $\tan \theta$

44:     If $\cos A = .28$ find the value of $\tan \dfrac{A}{2}$, and explain the resulting ambiguity.

45:     Find the values of (a). $\sin 7\dfrac{1}{2}^{\circ}$ (b). $\cos 7\dfrac{1}{2}^{\circ}$ (c). $\tan 22\dfrac{1}{2}^{\circ}$ and

(d). $\tan 11\dfrac{1}{4}^{\circ}$

46:     If $\sin \theta + \sin \phi = a$ and $\cos \theta + \cos \phi = b$, find the value of $\tan \dfrac{\theta - \phi}{2}$

Prove the following:

47:     $(\cos \alpha + \cos \beta)^2 + (\sin \alpha - \sin \beta)^2 = 4\cos^2 \dfrac{\alpha + \beta}{2}$

48:     $(\cos \alpha + \cos \beta)^2 + (\sin \alpha + \sin \beta)^2 = 4\cos^2 \dfrac{\alpha - \beta}{2}$

49:     $(\cos \alpha - \cos \beta)^2 + (\sin \alpha - \sin \beta)^2 = 4\sin^2 \dfrac{\alpha - \beta}{2}$

50:     $\sin A = \dfrac{2 \tan \dfrac{A}{2}}{1 + \tan^2 \dfrac{A}{2}}$

51: $\quad \cos A = \dfrac{1 - 2\tan^2 \dfrac{A}{2}}{1 + \tan^2 \dfrac{A}{2}}$

52: $\quad \sec\left(\dfrac{\pi}{4} + \theta\right)\sec\left(\dfrac{\pi}{4} - \theta\right) = 2\sec 2\theta$

53: $\quad \tan\left(\dfrac{\pi}{4} + \dfrac{A}{2}\right) = \sqrt{\dfrac{1 + \sin A}{1 - \sin A}} = \sec A + \tan A$

54: $\quad \sin^2\left(\dfrac{\pi}{8} + \dfrac{A}{2}\right) - \sin^2\left(\dfrac{\pi}{8} - \dfrac{A}{2}\right) = \dfrac{1}{\sqrt{2}}\sin A$

55: $\quad \cos^2 \alpha + \cos^2\left(\alpha + \dfrac{2\pi}{3}\right) + \cos^2\left(\alpha - \dfrac{2\pi}{3}\right) = \dfrac{3}{2}$

56: $\quad \cos^4 \dfrac{\pi}{8} + \cos^4 \dfrac{3\pi}{8} + \cos^4 \dfrac{5\pi}{8} + \cos^4 \dfrac{7\pi}{8} = \dfrac{3}{2}$

57: $\quad \sin^4 \dfrac{\pi}{8} + \sin^4 \dfrac{3\pi}{8} + \sin^4 \dfrac{5\pi}{8} + \sin^4 \dfrac{7\pi}{8} = \dfrac{3}{2}$

58: $\quad \cos 2\theta \cos 2\phi + \sin^2(\theta - \phi) - \sin^2(\theta + \phi) = \cos(2\theta + 2\phi)$

59: $\quad (\tan 4A + \tan 2A)(1 - \tan^2 3A \tan^2 A) = 2\tan 3A \sec^2 A$

60: $\quad \left(1 - \tan\dfrac{\alpha}{2} - \sec\dfrac{\alpha}{2}\right)\left(1 + \tan\dfrac{\alpha}{2} + \sec\dfrac{\alpha}{2}\right) + \sin \alpha \sec^2 \dfrac{\alpha}{2}$

Find the proper signs to be applied to the radicals in the following formulae:

61: $\quad 2\sin\dfrac{A}{2} = \pm\sqrt{1 - \sin A} \pm \sqrt{1 + \sin A}$, when $\dfrac{A}{2} = \dfrac{19\pi}{11}$

62: $\quad 2\cos\dfrac{A}{2} = \pm\sqrt{1 - \sin A} \pm \sqrt{1 + \sin A}$, when $\dfrac{A}{2} = -140°$

63: $\quad$ If $A = 340°$, prove that

$$2\sin\frac{A}{2} = -\sqrt{1+\sin A} + \sqrt{1-\sin A}$$

$$2\cos\frac{A}{2} = -\sqrt{1+\sin A} - \sqrt{1-\sin A}$$

64: If $A = 460°$, prove that

$$2\cos\frac{A}{2} = -\sqrt{1+\sin A} + \sqrt{1-\sin A}$$

65: If $A = 580°$ prove that $2\sin\frac{A}{2} = -\sqrt{1+\sin A} + \sqrt{1-\sin A}$

Within what respective limits must $\dfrac{A}{2}$ **lie when**

66: $2\sin\dfrac{A}{2} = \sqrt{1+\sin A} + \sqrt{1-\sin A}$

67: $2\sin\dfrac{A}{2} = -\sqrt{1+\sin A} + \sqrt{1-\sin A}$

68: $2\sin\dfrac{A}{2} = -\sqrt{1+\sin A} + \sqrt{1-\sin A}$

69: $2\cos\dfrac{A}{2} = \sqrt{1+\sin A} - \sqrt{1-\sin A}$

In the formula, $2\cos\dfrac{A}{2} = \pm\sqrt{1+\sin A} \pm \sqrt{1-\sin A}$ **find within**

what limits $\dfrac{A}{2}$ **must lie when**

70: The two positive signs are taken
71: The two negative signs are taken
72: The first sign is negative and the second positive.
73: Prove that the sine is algebraically less than the cosine for any
angle between $2n\pi - \dfrac{3\pi}{4}$ and $2n\pi + \dfrac{\pi}{4}$, where n is any integer.

74: If $\sin\dfrac{A}{3}$ be determined from the equation

$\sin A = 3\sin\dfrac{A}{3} - 4\sin^3\dfrac{A}{3}$, prove that we should expect to obtain also

the values of $\sin\dfrac{\pi - A}{3}$ and $-\sin\dfrac{\pi + A}{3}$, give also a geometrical illustration.

75: If $\cos\dfrac{A}{3}$ be found from the equation.

$\cos A = 4\cos^3\dfrac{A}{3} - 3\cos\dfrac{A}{3}$, prove that we should expect to obtain

also the values of $\cos\dfrac{2\pi - A}{3}$ and $\cos\dfrac{2\pi + A}{3}$

If $A + B + C = \pi$, prove that:

76: $\quad \sin 2A + \sin 2B - \sin 2C = 4\cos A\cos B\sin C$

77: $\quad \cos 2A + \cos 2B + \cos 2C = -1 - 4\cos A\cos B\cos C$

78: $\quad \cos 2A + \cos 2B - \cos 2C = 1 - 4\sin A\sin B\cos C$

79: $\quad \sin A + \sin B - \sin C = 4\sin\dfrac{A}{2}\sin\dfrac{B}{2}\cos\dfrac{C}{2}$

80: $\quad \cos A + \cos B + \cos C = 1 + 4\sin\dfrac{A}{2}\sin\dfrac{B}{2}\sin\dfrac{C}{2}$

81: $\quad \sin^2 A + \sin^2 B - \sin^2 C = 2\sin A\sin B\cos C$

82: $\quad \cos^2 A + \cos^2 B + \cos^2 C = 1 - 2\cos A\cos B\cos C$

83: $\quad \cos^2 A + \cos^2 B - \cos^2 C = 1 - 2\sin A\sin B\cos C$

84: $\quad \sin^2\dfrac{A}{2} + \sin^2\dfrac{B}{2} + \sin^2\dfrac{C}{2} = 1 - 2\sin\dfrac{A}{2}\sin\dfrac{B}{2}\sin\dfrac{C}{2}$

85: $\quad \sin^2\dfrac{A}{2} + \sin^2\dfrac{B}{2} - \sin^2\dfrac{C}{2} = 1 - 2\cos\dfrac{A}{2}\cos\dfrac{B}{2}\sin\dfrac{C}{2}$

Plane Trigonometry

86: $\tan\dfrac{A}{2}\tan\dfrac{B}{2}+\tan\dfrac{B}{2}\tan\dfrac{C}{2}+\tan\dfrac{C}{2}\tan\dfrac{A}{2}=1$

87: $\cot\dfrac{A}{2}+\cot\dfrac{B}{2}+\cot\dfrac{C}{2}=\cot\dfrac{A}{2}\cot\dfrac{B}{2}\cot\dfrac{C}{2}$

88: $\cot B\cot C+\cot C\cot A+\cot A\cot B=1$

89: $\sin(B+2C)+\sin(C+2A)+\sin(A+2B)$
$$=4\sin\dfrac{B-C}{2}\sin\dfrac{C-A}{2}\sin\dfrac{A-B}{2}$$

90: $\sin\dfrac{A}{2}+\sin\dfrac{B}{2}+\sin\dfrac{C}{2}-1=4\sin\dfrac{\pi-A}{4}\sin\dfrac{\pi-B}{4}\sin\dfrac{\pi-C}{4}$

91: $\cos\dfrac{A}{2}+\cos\dfrac{B}{2}-\cos\dfrac{C}{2}=4\cos\dfrac{\pi+A}{4}\cos\dfrac{\pi+B}{4}\cos\dfrac{\pi-C}{4}$

92: $\dfrac{\sin 2A+\sin 2B+\sin 2C}{\sin A+\sin B+\sin C}=8\sin\dfrac{A}{2}\sin\dfrac{B}{2}\sin\dfrac{C}{2}$

93: $\sin(B+C-A)+\sin(C+A-B)+\sin(A+B-C)$
$$=4\sin A\sin B\sin C$$

If A+B+C=2S, prove that:

94: $\sin(S-A)\sin(S-B)\sin S\sin(S-C)=\sin A\sin B$

95: $4\sin S\sin(S-A)\sin(S-B)\sin(S-C)$
$$=1-\cos^2 A-\cos^2 B-\cos^2 C+2\cos A\cos B\cos C$$

96: $\sin(S-A)+\sin(S-B)+\sin(S-C)-\sin S$
$$=4\sin\dfrac{A}{2}\sin\dfrac{B}{2}\sin\dfrac{C}{2}$$

97: $\cos^2 S+\cos^2(S-A)+\cos^2(S-B)+\cos^2(S-C)$
$$=2+2\cos A\cos B\cos C$$

98: If A+B+C=28, prove that

118

$$\cos^2 A + \cos^2 B + \cos^2 C + 2\cos A \cos B \cos C$$
$$= 1 + 4\cos S \cos(S-A)\cos(S-B)\cos(S-C)$$

99: If $\alpha + \beta + \gamma + \delta = 2\pi$, prove that

a.
$$\cos\alpha + \cos\beta + \cos\gamma + \cos\delta$$
$$+ 4\cos\frac{\alpha+\beta}{2}\cos\frac{\alpha+\gamma}{2}\cos\frac{\alpha+\delta}{2} = 0$$

b.
$$\sin\alpha - \sin\beta + \sin\gamma - \sin\delta$$
$$+ 4\cos\frac{\alpha+\beta}{2}\sin\frac{\alpha+\gamma}{2}\cos\frac{\alpha+\delta}{2} = 0$$

c.
$$\tan\alpha + \tan\beta + \tan\gamma + \tan\delta$$
$$= \tan\alpha\tan\beta\tan\gamma\tan\delta(\cot\alpha + \cot\beta + \cot\gamma + \cot\delta)$$

100: If the sum of four angles be 180°, prove that the sum of the products of their cosines taken two and two together is equal to the sum of the products of their sines taken similarly.

101: If $\alpha + \beta + \gamma = 0$, Prove that:
$$\sin 2\alpha + \sin 2\beta + \sin 2\gamma =$$
$$2(\sin\alpha + \sin\beta + \sin\gamma)(1 + \cos\alpha + \cos\beta + \cos\gamma)$$

102: Verify that:
$$\sin^3 a\sin(b-c) + \sin^3 b\sin(c-a) + \sin^3 c(a-b)$$
$$+ \sin(a+b+c)\sin(b-c)\sin(c-a)\sin(a-b) = 0$$

If A, B, C and D be any angles, prove that:

103:
$$\sin A \sin B \sin(A-B) + \sin B \sin C \sin(B-C)$$
$$+ \sin C \sin A \sin(C-A) + \sin(A-B)\sin(B-C)\sin(C-A)$$
$$= 0$$

104:
$$\sin(A-B)\cos(A+B) + \sin(B-C)\cos(B+C)$$
$$+ \sin(C-D)\cos(C+D) + \sin(D-A)\cos(D+A) = 0$$

105:
$$\sin(A+B-2C)\cos B - \sin(A+C-2B)\cos C = \sin(B-C)$$
$$\times\{\cos(B+C-A) + \cos(C+A-B) + \cos(A+B-C)\}$$

106:
$$\sin(A+B+C+D)+\sin(A+B-C-D)$$
$$+\sin(A+B-C+D)+\sin(A+B+C-D)$$
$$=4\sin(A+B)\cos C\cos D$$

107: If any theorem be true for values of A, B and C such that $A+B+C=180°$, Prove that the theorem is still true if we substitute for A, B and C respectively the quantities

    a. $90°-\dfrac{A}{2},90°-\dfrac{B}{2},90°-\dfrac{C}{2}$

    b. $180°-2A,180°-2B,180°-2C$

108: If $x+y+z=xyz$, prove that

$$\frac{3x-x^3}{1-3x^3}+\frac{3y-y^3}{1-3y^2}+\frac{3z-z^3}{1-3z^2}=\frac{3x-x^3}{1-3x^2}\frac{3y-y^3}{1-3y^2}\frac{3z-z^3}{1-3z^2}$$

109: If $x+y+z=xyz$, prove that

$$x(1-y^2)(1-z^2)+y(1-z^2)(1-x^2)+z(1-x^2)(1-y^2)=4xyz$$

13.10 Solutions of triangles

In a triangle:

1: Given $a = 25, b = 52$, and $c = 63$, find $\tan\dfrac{A}{2}, \tan\dfrac{B}{2}$ and $\tan\dfrac{C}{2}$

2: Given $a = 18, b = 24$ and $c = 30$ find $\sin A, \sin B$ and $\sin C$

3: Given $a = 35, b = 84$ and $c = 91$, find $\tan A, \tan B$ and $\tan C$

4: Given $a = 13, b = 14$ and $c = 15$, find the sines of the angles.

5: Given $a = \sqrt{3}, b = \sqrt{2}$, and $c = \dfrac{\sqrt{6}+\sqrt{2}}{2}$, find the angles.

In any triangle ABC, prove that

6: $\sin\dfrac{B-C}{2} = \dfrac{b-c}{a}\cos\dfrac{A}{2}$

7: $a(b\cos C - c\cos B) = b^2 - c^2$

8: $(b+c)\cos A + (c+a)\cos B + (a+b)\cos C = a+b+c$

9: $a(\cos B + \cos C) = 2(b+c)\sin^2\dfrac{A}{2}$

10: $a(\cos C - \cos B) = 2(b-c)\cos^2\dfrac{A}{2}$

11: $\dfrac{\sin(B-C)}{\sin(B+C)} = \dfrac{b^2 - c^2}{a^2}$

12: $\dfrac{a+b}{a-b} = \tan\dfrac{A+B}{2}\cot\dfrac{A-B}{2}$

13: $a\sin\left(\dfrac{A}{2}+B\right) = (b+c)\sin\dfrac{A}{2}$

In any triangle ABC, prove that

14:
$$\frac{a^2 \sin(B-C)}{\sin B + \sin C} + \frac{b^2 \sin(C-A)}{\sin C + \sin A} + \frac{c^2 \sin(A-B)}{\sin A + \sin B} = 0$$

15:
$$(b+c-a)\left(\cot\frac{B}{2} + \cot\frac{C}{2}\right) = 2a \cot\frac{A}{2}$$

16:
$$(a^2 - b^2 + c^2)\tan B = (a^2 + b^2 - c^2)\tan C$$

17:
$$c^2 = (a-b)^2 \cos^2\frac{C}{2} + (a+b)^2 \sin^2\frac{C}{2}$$

18:
$$a \sin(B-C) + b \sin(C-A) + c \sin(A-B) = 0$$

19:
$$\frac{a \sin(B-C)}{b^2 - c^2} = \frac{b \sin(C-A)}{c^2 - a^2} = \frac{c \sin(A-B)}{a^2 - b^2}$$

20:
$$a \sin\frac{A}{2}\sin\frac{B-C}{2} + b \sin\frac{B}{2}\sin\frac{C-A}{2} + c \sin\frac{C}{2}\sin\frac{A-B}{2} = 0$$

21:
$$a^2(\cos^2 B - \cos^2 C) + b^2(\cos^2 C - \cos^2 A)$$
$$+ c^2(\cos^2 A - \cos^2 B) = 0$$

22:
$$\frac{b^2 - c^2}{a^2}\sin 2A + \frac{c^2 - a^2}{b^2}\sin 2B + \frac{a^2 - b^2}{c^2}\sin 2C = 0$$

23:
$$\frac{(a+b+c)^2}{a^2 + b^2 + c^2} = \frac{\cot\frac{A}{2} + \cot\frac{B}{2} + \cot\frac{C}{2}}{\cot A + \cot B + \cot C}$$

24:
$$a^3 \cos(B-C) + b^3 \cos(C-A) + c^3 \cos(A-B) = 3abc$$

25: The sides of a right-angled triangle are 21 and 28 cm; find the length of the perpendicular drawn to the hypotenuse from the right angle.

26: If in any triangle the angles be to one another as 1:2:3, prove that the corresponding sides are $1 : \sqrt{3} : 2$.

27: In any triangle, if $\tan\frac{A}{2} = \frac{5}{6}$ and $\tan\frac{B}{2} = \frac{20}{37}$, find $\tan\frac{C}{2}$, and prove that in this triangle $a + c = 2b$.

28:     In an isosceles right-angled triangle a straight line is drawn from the middle point of one of the equal sides to the opposite angle. Show that it divides the angle into parts whose cotangents are 2 and 3.

29:     The perpendicular AD to the base of a triangle ABC divides it into segments such that BD, CD and AD are in the ratio of 2, 3 and 6; prove that the vertical angle of the triangle is $45°$.

30:     A ring, ten cm in diameter, is suspended from a point 12 cm above its centre by 6 equal strings attached to its circumference at equal intervals. Find the cosine of the angle between consecutive strings.

31:     If a^2, b^2, and c^2 be in A.P., prove that $\cot A, \cot B$ and $\cot C$ are in A.P., also.

32:     If a, b and c be in A.P., prove that $\cos A \cot \dfrac{A}{2}$, $\cos B \cot \dfrac{B}{2}$ and $\cos C \cot \dfrac{C}{2}$ are in A.P.

33:     If a, b and c are in H.P., prove that $\sin^2 \dfrac{A}{2}, \sin^2 \dfrac{B}{2}$ and $\sin^2 \dfrac{C}{2}$ are also in H.P.

34:     The sides of a triangle are in A.P. and the greatest and least angles are θ and ϕ. Prove that: $4(1-\cos\theta)(1-\cos\phi) = \cos\theta + \cos\phi$

35:     The sides of a triangle are in A.P. and the greatest angles exceeds the least by $90°$; provide that the sides are proportional to $\sqrt{7+1}, \sqrt{7}$, and $\sqrt{7-1}$.

36:     If $C = 60°$, then prove that $\dfrac{1}{a+c} + \dfrac{1}{b+c} = \dfrac{3}{a+b+c}$

37:     In any triangle ABC if D be any point of the base BC, such that BD:DC $:: m:n$, and if $\log 25.784 = 1.4113503$, $\lfloor DAC = \beta$, $\lfloor CDA = \theta$,   and $AD = x$,   prove   that $(m+n)\cot\theta = m\cot\alpha - n\cot\beta = n\cot B - m\cot C$,

$(m+n)^2 x^2 = (m+n)(mb^2 + nc^2) - mn\alpha^2$

38:     If in a triangle the bisector of the side c be perpendicular to the side b, prove that $2\tan A + \tan C = 0$

39:    In any triangle prove that, if θ be any angle, then
$b\cos\theta = c\cos(A-\theta) + a\cos(C+\theta)$

40:    If p and q be the perpendiculars from the angular points A and B on any line passing through the vertex C of the triangle ABC, then prove that $a^2 p^2 + b^2 q^2 - 2abpq\cos C = a^2 b^2 \sin^2 C$

41:    In the triangle ABC, line OA, OB and OC are drawn so that the angles OAB, OBC and OCA are each equal to ω; prove that
$\cot\omega = \cot A + \cot B + \cot C$ and
$\operatorname{cosec}^2 \omega = \operatorname{cosec}^2 A + \operatorname{cosec}^2 B + \operatorname{cosec}^2 C$

13.11 Heights and distances

1:     A person, standing on the bank of a river, observes that the angle subtended by a tree on the opposite bank is 60°; when he retires 20 meters from the bank, he finds the angle to be 30°; find the height of the tree and the breadth of the river.

2:     At a certain point the angle of elevation of a tower is found to be such that its cotangent is $\frac{3}{5}$; on walking 32 meters directly toward the tower its angle of elevation is an angle whose cotangent is $\frac{2}{5}$. Find the height of the tower.

3:     At a point A, the angle of elevation of a tower is found to be such that its tangent is $\frac{5}{12}$; on walking 240 meters nearer the tower the tangent of the angle of elevation is found to be $\frac{3}{4}$; what is the height of the tower?

4:     Find the height of a chimney when it is found that, on walking towards it 50 meters in a horizontal line through its base, the angular elevation of its top changes from 30° to 45°.

5:     An observer on the top of a cliff, 200 meters above the sea-level, observes the angles of depression of two ships at anchor to be 45° and 30° respectively; find the distances between the ships if the line joining them points to the base of the cliff.

6:     From the top of a cliff an observer finds that the angles of depression of two buoys in the sea are 39° and 26° respectively; the buoys are 300 meters apart and the line joining them points straight at the foot of the cliff; find the height of the cliff and the distance of the nearest buoy from the foot of the cliff, given that cot 26° = 2.0503 , and cot 39° = 1.2349.

7:     The upper part of a tree broken over by the wind makes an angle of 30° with the ground, and the distance from the root to the point where the top of the tree touches the ground is 10 m., what was the height of the tree?

8:     The horizontal distance between two towers is 60 meters and the angular depression of the top of the first as seen from the top of the second, which is 150 m high, is 30°; find the height of the first.

9:     The angle of elevation of the top of an unfinished tower at a point distant 120 meters from its base is 45°; how much higher must the tower be raised so that its angle of elevation at the same point may be 60°?

10:     Two pillars of equal height stand on either side of a wide road which is 100 meters wide; at a point in the road between the towers the elevation of the tops of the towers are 60° and 30°; find their height and the position of the point.

11:     The angle of elevation of the top of a tower is observed to be 60°; at a point 40 meter above the first point of observation the elevation is found to be 45°; find the height of the tower and its horizontal distance from the points of observation.

12:     At the foot of a mountain the elevation of its summit is found to be 45°; after ascending 1000 m towards the mountain up a slope of 30° inclination the elevation is found to be 60°. Find the height of the mountain.

13:     What is the angle of elevation of the sun when the length of the shadow of a pole is $\sqrt{3}$ times the height of the pole?

14:     The shadow of a tower standing on a level plane is found to be 60 meters longer when the sun's altitude is 30° than when it is 45°. Prove that the height of the tower is $30(1+\sqrt{3})$ meters.

15:     On a straight coast there are three objects A, B and C, such that AB = BC = 2 Km. A vessel approaches B in a line perpendicular to the coast, and at a certain point AC is found to subtend an angle of 60°; after sailing in the same direction for ten minutes AC is found to subtend an angle of 120°; find the rate at which the ship is going.

16:     Two flagstaffs stand on a horizontal plane. A and B are two points on the line joining the bases of the flagstaffs and between them. The angles of elevation of the tops of the flagstaffs as seen from A are 30° and 60° and, as seen from B, they are 60° and 45°. If the length AB be 10 meters, find the heights of the flagstaffs and the distance between them.

17: P is the top and Q the foot of a tower standing on a horizontal plane. A and B are two points on this plane such that AB is 32 m and QAB is a right angle. It is found that $\cot PAQ = \dfrac{2}{5}$, and $\cot PBQ = \dfrac{3}{5}$. Find the height of the tower.

18: A light house, facing north, sends out a fan-shaped beam of light extending from north-east to north-west. An observer on a steamer, sailing due west, first sees the light when he is 5 Km away from the lighthouse and continues to see it for $30\sqrt{2}$ minutes. What is the speed of the steamer?

19: A man stands at a point X on the bank XY of a river with straight and parallel banks, and observes that the line joining X to a point Z on the opposite bank makes an angle of 30° with XY. He then goes along the bank a distance of 200 meters to Y and finds that the angle Z Y X is 60°. Find the breadth of the river.

20: A man, walking due north, observes that the elevation of a balloon, which is due east of him and is sailing toward the north-west, is then 60°, after he has walked 400 meters the balloon is vertically over his head; find its height supposing it to have always remained the same.

Solve the following:

21: A flagstaff stands on the middle of a square tower. A man on the ground, opposite the middle of one face and distant from it 100 meters, just sees the flag; on his receding another 100 feet, the tangents of elevation of the top of the tower and the top of the flagstaff are found to be $\dfrac{1}{2}$ and $\dfrac{5}{9}$. Find the dimensions of the tower and the height of the flagstaff, the ground being horizontal.

22: A man, walking on a level plane towards a tower, observes that at a certain point the angular height of the tower is $10°$, and, after going 50 meters nearer the tower, the elevation is found to be $15°$. Having given $\angle \sin 15° = 9.4129962$, $\angle \cos 5° = 9.9983442$,

$\log 25.783 = 1.4113334$, and $\log 25.784 = 1.4113503$, find to 4 places of decimals the height of the tower in meters.

23:    DE is a tower standing on a horizontal plan and ABCD is a straight line in the plane. The height of the tower subtends an angle θ at A, 2θ at B, and 3θ at C. if AB and BC be respectively 50 and 20 meters, find the height of the tower and the distance CD.

24:    A tower, 50 meters high, stands on the top of a mound; from a point on the ground the angles of elevation of the top and bottom of the tower are found to be 75° and 45° respectively; find the height of the mound.

25:    A vertical pole (more than 10 meters high) consists of two parts, the lower being $\frac{1}{3}$ rd of the whole. From a point in a horizontal plan through the foot of the pole and 40 meters from it, the upper part subtends an angle whose tangent is $\frac{1}{2}$. Find the height of the pole.

26:    A tower subtends an angle α at a point on the same level as the foot of the tower, and at a second point, h meters above the first, the depression of the foot of the tower is β Find the height of the tower.

27:    A person in a balloon, which has ascended vertically from flat land at the sea-level, observes the angle of depression of a ship at anchor to be 30°; after descending vertically for 600 meters, he finds the angle of depression to be 15°; find the horizontal distance of the ship from the point of ascent.

28:    PQ is a tower standing on a horizontal plan, Q being its foot; A and B are the two points on the plane such that the $\angle$ QAB is 90°, and AB is 40 meters. It is found that $\cot PAQ = \frac{3}{10}$ and $\cot PBQ = \frac{1}{2}$. Find the height of the tower.

29:    A column is E.S.E of an observer, and at noon the end of the shadow is North-East of him. The shadow is 80 meters long and the elevation of the column at the observer's station is 45°. Find the height of the column.

30: A tower is observed from two stations A and B. It is found to be due north of A and north-west of B. B is due east of A and distant from it 10 meters. The elevation of the tower as seen from A is the complement of the elevation as seen from B. Find the height of the tower.

31: The elevation of the steeple at a place due south of it is $45°$ and at another place due west of the former place the elevation is $15°$. If the distance between the two places be a, prove that the height of the steeple is $\dfrac{a(\sqrt{3}-1)}{24\sqrt{3}}$

32: A person stands in a diagonal produced of the square base of a church tower, at a distance 2a from it, and observes the angles of elevation of each of the two outer corners of the top of the tower to be $30°$, whilst that of the nearest corner is $45°$. Prove that the breadth of the tower is $(\sqrt{10}-\sqrt{2})$.

33: A person standing at a point A due south of a tower built on a horizontal plane observes the altitude of the tower to be $60°$. He then walks to B due west of A and observes the altitude to be $45°$, and again at C in AB produced he observes it to be $30°$. Prove that B is midway between A and C.

34: At each end of a horizontal base of length 2a it is found that the angular height of a certain peak is θ and that at the middle point it is φ. Prove that the vertical height of the peak is

$$\frac{a\sin\theta\sin\varphi}{\sqrt{\sin(\varphi+\theta)\sin(\varphi-\theta)}}$$

35: A and B are two stations 1000 m apart; P and Q are two stations in the same plane as AB and on the same side of it; the angles PAB, PBA, QAB and QBA are respectively $75°$, $30°$, $45°$, and $90°$; find how far P is from Q and how far each is from A and B. For the following seven examples the required logarithms must be taken from the tables.

36: At a point on a horizontal plane the elevation of the summit of a mountain is found to be $22°15'$, and at another point on the plane, 1 km farther away in a direct line, its elevation is $10°12'$; find the height of the mountain.

37: From the top of a hill the angles of depression of two successive points, 1 km apart on level ground and in the same vertical plane with the observe, are found to be $5°$ and $10°$ respectively. Find the height of the hill and the horizontal distance to the nearest point.

38: A cliff and a tower stand on the same horizontal plane. The height of the tower is 140 meters, and the angles of depression of the top and bottom of the tower as seen from the top of the cliff are $40°$ and $80°$ respectively. Find the height of the tower.

39: A tower PN stands on level ground. A base AB is measured at right angles to AN, the points A, B, and N being in the same horizontal plane, and the angles PAN and PBN are found to be α and β respectively. Prove that the height of the tower is

$$AB\frac{\sin\alpha\sin\beta}{\sqrt{\sin(\alpha-\beta)\sin(\alpha+\beta)}}$$ If AB = 100 m, $\alpha=70°$ and $\beta=50°$, calculate the height.

40: A man, standing due south of a tower on a horizontal plane through its foot, finds the elevation of the top of the tower to be $54°16'$; he goes east 100 meters and finds the elevation to be then $50°8'$. Find the height of the tower.

41: A man in a balloon observes that the angle of depression of an object on the ground bearing due north is $33°$; the balloon drifts 3 miles due west and the angle of depression is now found to be $21°$. Find the height of the balloon.

42: From the extremities of a horizontal base-line AB, whose length is 1 km, the bearings of the foot C of a tower are observed and it is found that $\angle CAB = 56°23'$, $\angle CBA = 47°15'$, and the elevation of the tower from A is $9°25'$; find the height of the tower.

Solve the following:

43: A bridge has 5 equal spans, each of 10 m measured from the centre of the piers, and a boat is moored in a line with one of the middle piers. The whole length of the bridge subtends a right angle as seen from the boat. Prove that the distance of the boat from the bridge is $10\sqrt{6}$ m.

44: A ladder placed at an angle of $75°$ with the ground just reaches the sill of a window at a height of 9 meters above the ground on one side of a street. On turning the ladder over without moving its foot, it is found that when it rests against a wall on the other side of the street it is at an angle of $15°$ with the ground. Prove that the breadth of the street and the length of the ladder are respectively $9(3-\sqrt{3})$ and $9(\sqrt{6}-\sqrt{2})$ meters.

45: From a house on one side of a street observations are made of the angle subtended by the height of the opposite house; from the level of the street, the angle subtended is the angle whose tangent is 3; from two windows one above the other the angle subtended is found to be the angle whose tangent is -3; the height of the opposite house being 30 meters, find the height above the street of each of the two windows.

46: A rod of given length can turn in a vertical plane passing through the sun, one end being fixed on the ground; find the longest shadow it can cast on the ground. Calculate the altitude of the sun when the longest shadow it can cast is $3\dfrac{1}{2}$ times the length of the rod.

47: A person on a ship A observes another ship B leaving a harbor, whose bearing is then N.W. After 10 minutes A, having sailed one km. N.E., sees B due west and the harbor then bears $60°$ west of north. After another 10 minutes B is observed to bear S.W. Find the distances between A and B at the first observation and also the direction and rate of B.

48: A person on a ship sailing north sees two lighthouses, which are 6 km. apart, in a line due west; after an hour's sailing one of them bears S.W and the other S.S.W. Find the ship's rate.

49: A person on a ship sees a lighthouse N.W. of himself. After sailing for 12 Km. in a direction $15°$ south of W. the lighthouse is seen due N. Find the distance of the lighthouse from the ship in each position.

50:     A man, travelling west along a straight road, observes that when he is due south of a certain windmill the straight line drawn to a distant tower makes an angle of $30°$ with the road. One Km. farther on the bearings of the windmill and tower are respectively N.E. and N.W. Find the distances of the tower from the windmill and from the nearest point of the road.

51:     An observer on a headland sees a ship due north of him; after a quarter of an hour he sees it due east and after another half-hour he sees it due south-east; find the angle that the ship's course makes with the meridian and the time after the ship is first seen until it is nearest the observer, supposing that it sails uniformly in a straight line.

52:     A man walking along a straight road, which runs in a direction $30°$ east of north, notes when he is due south of a certain house; when he has walked 1 km. farther, he observes that the house lies due west and that a windmill on the opposite side of the road is N.E. of him; three km farther on he finds that he is due north of the windmill; prove that the line joining the house and the windmill makes with the road the angle whose tangent is $\dfrac{48-25\sqrt{3}}{11}$

53:     A, B and C are three consecutive milestones on a straight road from each of which a distant spire is visible. The spire is observed to bear north-east of A, east at B, and $60°$ east of south at C. Prove that the shortest distance of the spire from the road is $\dfrac{7+5\sqrt{3}}{13}$ km.

54:     Two stations due south of a tower, which leans towards the north, are at distances a and b from its foot; and α and β are the elevations of the top of the tower from these stations, prove that its inclination to the horizontal is $\cot^{-1}\dfrac{b\cot\alpha-a\cot\beta}{b-a}$

55:     From a point A on a level plane the angle of elevation of a balloon is α, the balloon being south of A; from a point B, which is at a distance c south of A, the balloon is seen northwards at an elevation of β; find the distance of the balloon from A and its height above the ground.

56:    A statue on the top of a pillar subtends the same angle α at distances of 9 and 11 m from the pillar; if $\tan\alpha = \dfrac{1}{10}$, find the height of the pillar and of the statue.

57:    A flagstaff on the top of a tower is observed to subtend the same angle α at two points on a horizontal plane, which lie on a line passing through the centre of the base of the tower and whose distance from one another is $2a$, and an angle β at a point halfway between them. Prove that the height of the flagstaff is

$$a\sin\alpha\sqrt{\frac{2\sin\beta}{\cos\alpha\sin(\beta-\alpha)}}$$

58:    An observer in the first place stations himself at a distance a meters from a column standing upon a mound. He finds that the column subtends an angle, whose tangent is $\dfrac{1}{2}$, at his eye which may be supposed to be on the horizontal plane through the base of the mound. On moving $\dfrac{2}{3}a$ m nearer the column, he finds that the angle subtended is unchanged. Find the height of the mound and of the column.

59:    A church tower stands on the bank of a river, which is 50 m. wide, and on the top of the tower is a spire 10 m. high. To an observer on the opposite bank of the river, the spire subtends the same angle that a pole 2 m. high subtends when placed upright on the ground at the foot of the tower. Prove that the height of the tower is nearly 95 m.

60:    A person, wishing to ascertain the height of a tower, stations himself on a horizontal plane through its foot at a point at which the elevation of the top in $30°$. On walking a distance a in a certain direction he finds that the elevation of the top is the same as before, and on then walking a distance $\dfrac{5}{3}a$ at right angles to his former direction he finds the elevation of the top to be $60°$. Prove that the height of the tower is either $\sqrt{\dfrac{5}{6}}a$ or $\sqrt{\dfrac{85}{48}}a$.

61:     The angles of elevation of the top of a tower, standing on a horizontal plane, from two points distant a m. and b m. from the base and in the same straight line with it are complementary. Prove that the height of the tower is $\sqrt{ab}$ m., and, if θ be the angle subtended at the top of the tower by the line joining the two points, then $\sin\theta = \dfrac{a \sim b}{a+b}$

62:     A tower 150 m. high stands on the top of a cliff 80 m. high. To an observer on a ship, the tower and cliff subtend equal angles. How far from the cliff is the observer if he is 5m. above the plane passing through the foot of the cliff?

63:     A statue on the top of a pillar, standing on level ground, is found to subtend the greatest angle a at the eye of an observer when his distance from the pillar is c m.; prove that the height of the statue is $2c \tan a$ m., and find the height of the pillar.

64:     A tower stood at the foot of an inclined plane whose inclination to the horizon was $9°$. A line 100 m. in length was measured straight up the incline from the foot of the tower, and at the end of this line the tower subtended an angle of $54°$. Find the height of the tower, having given $\log 2 = .30103$, $\log 114.4123 = 2.0584726$, and $\angle \sin 54° = 9.9079576$.

65:     A vertical tower stands on a declivity which is inclined at $15°$ to the horizon. From the foot of the tower a man ascends the declivity for 80 meters, and then finds that the tower subtends an angle of $30°$. Prove that the height of the tower is $40(\sqrt{6} - \sqrt{2})$ meters.

66:     The altitude of a certain rock is $47°$, and after walking towards it 1 km up a slope inclined at $30°$ to the horizon an observer finds its altitude to be $77°$. Find the vertical height of the rock above the first point of observation, given that $\sin 47° = .73135$.

67:     A man observes that, when he has walked c meters up an inclined plane, the angular depression of an object in a horizontal plane through the foot of the slope is α, and that, when he has walked a further distance of c feet, the depression is β. Prove that the inclination of the slope to the horizon is the angle whose cotangent is $(2\cot\beta - \cot\alpha)$.

68: A regular pyramid on a square base has an edge 150 m long, and the length of the side of its base is 200 m. Find the inclination of its face to the base.

69: A pyramid has for base a square of side a; its vertex lies on a line through the middle point of the base and perpendicular to it, and at a distance h from it; prove that the angle α between the two lateral faces is given by the equation $\sin\alpha = \dfrac{2h\sqrt{2a^2 + 4h^2}}{a^2 + 4h^2}$

70: A flagstaff, 10 meters high, stands in the centre of an equilateral triangle which is horizontal. From the top of the flagstaff each side subtends an angle of $60°$; prove that the length of the side of the triangle is $5\sqrt{6}$ meters.

71: The extremity of the shadow of a flagstaff, which is 6 m high and stands on the top of a pyramid on a square base, just reaches the side of the base and is distant 56 and 8 m respectively from the extremities of that side. Find the sun's altitude if the height of the pyramid be 34 m.

72: The extremity of the shadow of a flagstaff, which is 6 meters high and stands on the top of a pyramid on a square base, just reaches the side of the base and is distant x feet and y feet respectively from the ends of that side; prove that the height of the pyramid is

$\sqrt{\dfrac{x^2 + y^2}{2}}\tan\alpha - 6$, where α is the elevation of the sun.

73: The angle of elevation of a cloud from a point h meters above a lake is α, and the angle of depression of its reflex ion in the lake is β; prove that its height is
$h\dfrac{\sin(\beta + \alpha)}{\sin(\beta - \alpha)}$

74: The shadow of a tower is observed to be half the known height of the tower and sometime afterwards it is equal to the known height; how much will the sun have gone down in the interval, given $\log 2 = .30103$, $\angle\tan 63°26' = 10.3009994$ and diff. for 1' = 3159?

75: An isosceles triangle of wood is placed in a vertical plane, vertex upwards, and faces the sun. If 2a be the base of the triangle, h its height, and $30°$ the altitude of the sun, prove that the tangent of the angle   at   the   apex   of   the   shadow   is

$$\frac{2ab\sqrt{3}}{3b^2 - a^2}$$

76:     A rectangular target faces due south, being vertical and standing on a horizontal plane. Compare the area of the target with that of its shadow on the ground when the sun is $\beta°$ from the south at an altitude of $\alpha°$.

77:     A spherical ball, of diameter δ, subtends an angle α at a man's eye when the elevation of its centre is β; prove that the height of the centre of the ball is $\frac{1}{2}\delta \sin\beta \operatorname{cosec}\frac{\alpha}{2}$

78:     A man standing on a plane observes a row of equal and equidistant pillars, the 10th and 17th of which subtend the same angle that they would do if they were in the position of the first and were respectively $\frac{1}{2}$ and $\frac{1}{3}$ of their height. Prove that, neglecting the height of the man's eye, the line of pillars is inclined to the line drawn from his eye to the first at an angle whose secant is nearly 2.6 For the following nine examples the required logarithms must be taken from the tables.

79:     A and B are two points, which are on the banks of a river and opposite to one another, and between them is the mast, PN, of a ship; the breadth of the river is 100 meters, and the angular elevation of P at A is $14°20'$ and at B is $8°10'$. What is the height of P above AB?

80:     AB is a line 1000 meters long; B is due north of A and from B a distant point P bears $70°$ east of north; at a it bears $41°22'$ east of north; find the distance from A to P.

81:     A is a station exactly 10 km west of B. The bearing of a particular rock from A is $74°19'$ east of north, and its bearing from B is $26°51'$ west of north. How far is it north of the line AB?

82:     The summit of a spire is vertically over the middle point of a horizontal square enclosure whose side is of length a meters; the height of the spire is h meters above the level of the square. If the shadow of the spire just reaches a corner of the square when the sun has an altitude θ, prove that $h\sqrt{2} = a\tan\theta$. Calculate h, having given $a= 100$ meters and $\theta = 25°15'$.

83:    Walking along a straight level road in a direction N.W., I notice two spires, P and Q, in a straight line with me on a bearing N. 20° E., P being the nearer spire. After walking 4 km. farther along the road, P bears E. 22° S. and Q bears E. 26° N. Find the distance between the spires.

84:    AB is a road running uphill in a direction due east at an inclination of 10°, B being above A and at a distance of 500 meters from A. The bearing of an object c from A is E. 50° N. at an elevation of 43°, and the bearing of c from B is W. 65° N. calculate the horizontal and vertical distances of C from A.

85:    A wireless signal from an aero plane is intercepted at two direction finding stations, A and B, which are five km. apart in a north and south line. From A the direction of the aero plane is found to be 66° west of north, and from B it is found to be 20° west of south. At the same time the altitude of the aero plane is observed from A to be 40°. Find the height of the aero plane above A.

86:    An observer sees an aero plane due N. at an elevation of 8°. Two minutes later he sees it N.E. at the same elevation. It is known to be going due E., the horizontal component of its velocity being 80 km an hour. Show that it is rising at the rate of nearly 77.615 meters per minute.

87:    A small balloon is released from a point on a level plain, and ascends with a constant vertical velocity of 100 meters per minute. After it has risen for 10 minutes it is at P, and one minute later is at Q. At a station O on the plain the bearing of P is due east, and OP makes an angle of 63° with the plain. At O the bearing of Q is 29° north of east, and OQ makes an angle of 70° with the plain. Assuming the wind to blow with constant horizontal velocity, find this velocity, and show that its direction is about 50°37' north of west.

13.12 Inverse Circular ratios

Prove the following statements:

1: $\quad \sin^{-1}\dfrac{3}{5} + \sin^{-1}\dfrac{8}{17} = \sin^{-1}\dfrac{77}{85}$

2: $\quad \sin^{-1}\dfrac{5}{13} + \sin^{-1}\dfrac{7}{25} = \cos^{-1}\left(\dfrac{253}{325}\right)$

3: $\quad \cos^{-1}\dfrac{4}{5} + \cos^{-1}\dfrac{12}{13} = \cos^{-1}\dfrac{33}{65}$

4: $\quad \cos^{-1}x = 2\sin^{-1}\sqrt{\dfrac{1-x}{2}} = 2\cos^{-1}\sqrt{\dfrac{1+x}{2}}$

5: $\quad 2\cos^{-1}\dfrac{3}{\sqrt{13}} + \cot^{-1}\dfrac{16}{63} + \dfrac{1}{2}\cos^{-1}\dfrac{7}{25} = \pi$

6: $\quad \tan^{-1}\dfrac{1}{2} + \tan^{-1}\dfrac{1}{3} = \sin^{-1}\dfrac{1}{\sqrt{5}} + \cot^{-1}3 = 45°$

7: $\quad \tan^{-1}\dfrac{2}{3} = \dfrac{1}{2}\tan^{-1}\dfrac{12}{5}$

8: $\quad 2\tan^{-1}\dfrac{1}{5} + \tan^{-1}\dfrac{1}{7} + 2\tan^{-1}\dfrac{1}{8} = \dfrac{\pi}{4}$

9: $\quad \tan^{-1}\dfrac{3}{4} + \tan^{-1}\dfrac{3}{5} - \tan^{-1}\dfrac{8}{19} = \dfrac{\pi}{4}$

10: $\quad \tan^{-1}\dfrac{1}{3} + \tan^{-1}\dfrac{1}{5} + \tan^{-1}\dfrac{1}{7} + \tan^{-1}\dfrac{1}{8} = \dfrac{\pi}{4}$

11: $\quad 3\tan^{-1}\dfrac{1}{4} + \tan^{-1}\dfrac{1}{20} = \dfrac{\pi}{4} - \tan^{-1}\dfrac{1}{1985}$

12: $\quad 4\tan^{-1}\dfrac{1}{5} - \tan^{-1}\dfrac{1}{70} + \tan^{-1}\dfrac{1}{99} = \dfrac{\pi}{4}$

13: $\quad \tan^{-1}\dfrac{120}{119} = 2\sin^{-1}\dfrac{5}{13}$

14:   $\tan^{-1} t + \tan^{-1} \dfrac{2t}{1-t^2} = \tan^{-1} \dfrac{3t-t^3}{1-3t^2}$, t being   positive   if

$t < \dfrac{1}{\sqrt{3}}$ or $t > \sqrt{3}$ and $= \pi + \tan^{-1} \dfrac{3t-t^3}{1-3t^2}$ if $\sqrt{3} > t > \dfrac{1}{\sqrt{3}}$.

15:   $\tan^{-1} \sqrt{\dfrac{a(a+b+c)}{bc}} + \tan^{-1} \sqrt{\dfrac{b(a+b+c)}{ca}}$

$+ \tan^{-1} \sqrt{\dfrac{c(a+b+c)}{ab}} = \pi$

16:   $\cot^{-1} \dfrac{ab+1}{a-b} + \cot^{-1} \dfrac{bc+1}{b-c} + \cot^{-1} \dfrac{ca+1}{c-a} = 0$

17:   $\tan^{-1} n + \cot^{-1}(n+1) = \tan^{-1}(n^2+n+1)$

18:   $\cos\left(2\tan^{-1}\dfrac{1}{7}\right) = \sin\left(4\tan^{-1}\dfrac{1}{3}\right)$

19:   $2\tan^{-1}\left[\tan(45°-\alpha)\tan\dfrac{\beta}{2}\right] = \cos^{-1}\left[\dfrac{\sin 2\alpha + \cos\beta}{1+\sin 2\alpha \cos\beta}\right]$

20:   $2\tan^{-1}\left[\tan\dfrac{\alpha}{2}\tan\left(\dfrac{\pi}{4}-\dfrac{\beta}{2}\right)\right] = \tan^{-1}\dfrac{\sin\alpha\cos\beta}{\sin\beta + \cos\alpha}$

21:   Show that

$$\cos^{-1}\sqrt{\dfrac{a-x}{a-b}} = \sin^{-1}\sqrt{\dfrac{x-b}{a-b}} = \cot^{-1}\sqrt{\dfrac{a-x}{x-b}}$$

$$= \dfrac{1}{2}\sin^{-1}\dfrac{2\sqrt{(a-x)(x-b)}}{a-b}$$

22:   If $\cos^{-1}\dfrac{x}{a} + \cos^{-1}\dfrac{y}{b} = a$, prove that

$$\dfrac{x^2}{a^2} - \dfrac{2xy}{ab}\cos a + \dfrac{y^2}{b^2} = \sin^2 a$$

Solve the equations:

23: $\quad \tan^{-1}\dfrac{\sqrt{1+x^2}-\sqrt{1-x^2}}{\sqrt{1+x^2}+\sqrt{1-x^2}}=\beta$

24: $\quad \tan^{-1}2x+\tan^{-1}3x=\dfrac{\pi}{4}$

25: $\quad \tan^{-1}\dfrac{x-1}{x-2}+\tan^{-1}\dfrac{x+1}{x+2}=\dfrac{\pi}{4}$

26: $\quad \tan^{-1}(x+1)+\cot^{-1}(x-1)=\sin^{-1}\dfrac{4}{5}+\cos^{-1}\dfrac{3}{5}$

27: $\quad \tan^{-1}(x+1)+\tan^{-1}(x-1)=\tan^{-1}\dfrac{8}{31}$

28: $\quad 2\tan^{-1}(\cos x)=\tan^{-1}(2\ \mathrm{cosec}\,x)$

29: $\quad \tan^{-1}x+2\ \cot^{-1}x=\dfrac{2}{3}\pi$

30: $\quad \tan\cos^{-1}x=\sin\cot^{-1}\dfrac{1}{2}$

31: $\quad \cot^{-1}x-\cot^{-1}(x+2)=15°$

32: $\quad \cos^{-1}\dfrac{x^2-1}{x^2+1}+\tan^{-1}\dfrac{2x}{x^2-1}=\dfrac{2\pi}{3}$

33: $\quad \cot^{-1}x+\cot^{-1}(n^2-x+1)=\cot^{-1}(n-1)$

34: $\quad \sin^{-1}x+\sin^{-1}2x=\dfrac{\pi}{3}$

35: $\quad \sin^{-1}\dfrac{5}{x}+\sin^{-1}\dfrac{12}{x}=\dfrac{\pi}{2}$

36: $\quad \tan^{-1}\dfrac{a}{x}+\tan^{-1}\dfrac{b}{x}+\tan^{-1}\dfrac{c}{x}+\tan^{-1}\dfrac{d}{x}=\dfrac{\pi}{2}$

37: $\quad \sec^{-1}\dfrac{x}{a}-\sec^{-1}\dfrac{x}{b}=\sec^{-1}b-\sec^{-1}a$

38: $\quad \mathrm{cosec}^{-1}x=\mathrm{cosec}^{-1}a+\mathrm{cosec}^{-1}b$

39: $\quad 2\tan^{-1} x = \cos^{-1}\dfrac{1-a^2}{1+a^2} - \cos^{-1}\dfrac{1-b^2}{1+b^2}$

14 Closing Thoughts

Mathematics is not a spectator sport. The patterns and underlying nuances are like a work of art. The more you apply yourself to the subject, the more you uncover and understand. Mathematics is a subject which requires practice. This is not something that you relax on your couch, casually browse through and hope to achieve mastery. This will require patience and application.

1. There are five fundamental principles, or say **good habits** that we would like to emphasize the same, one more time.

2. Neatness is conducive to accuracy. Refrain from the temptation to write down something quickly and scratching the same to make the necessary corrections.

3. One of the weakness we find in student while solving word problems is the usage of = sign. This sign as a specific meaning in the world of mathematics. It cannot be used as a way to begin every new line of step in the problem solving process. Use appropriate mathematical signs and symbols. Never use them to mean something vague. = Sign is never good space filler.

4. Spend a second or two to explain how you arrived at a certain step. Several books and references use a statement, such as ``it follows from the above statement''. We have oftentimes wondered how the expression or equation below follows from the one above. A good explanation is an excellent demonstration of your understanding of the underlying principles.

5. When you are faced with several conclusions during a problem solving process, it is a good idea to number the statements or equations. In subsequent steps, you can refer to these conclusions by using the label or the assigned equation number.

6. The easiest of problems attracts the silliest of mistakes. If the problem is easy, motivate yourself to get it right. Do not let over-confidence or carelessness to take control of the situation.